BUS

D1121624

CERAMIC TECHNOLOGY
AND PROCESSING

CERAMIC TECHNOLOGY AND PROCESSING

by

Alan G. King

Twinsburg, Ohio

NOYES PUBLICATIONS

WILLIAM ANDREW PUBLISHING
Norwich, New York, U.S.A.

Library of Congress Catalog Card Number: 00-023980
ISBN: 0-8155-1443-3
Printed in the United States

Published in the United States of America by
Noyes Publications / William Andrew Publishing
13 Eaton Avenue
Norwich, NY 13815
1-800-932-7045
www.williamandrew.com
www.knovel.com

10 9 8 7 6 5 4 3 2 1

NOTICE

Library of Congress Cataloging-in-Publication Data

Ceramic technology and processing / by Alan G. King.
    p.   cm.
Includes bibliographical references and index.
ISBN 0-8155-1443-3
1. Ceramics. I. Title.
TP807.K49   2002   00-023980
666--dc21   CIP

MATERIALS SCIENCE AND PROCESS TECHNOLOGY SERIES

Series Editors

Gary E. McGuire, Microelectronics Center of North Carolina
Stephen M. Rossnagel, IBM Thomas J. Watson Research Center
Rointan F. Bunshah, University of California, Los Angeles (1927–1999), founding editor

Electronic Materials and Process Technology

CHARACTERIZATION OF SEMICONDUCTOR MATERIALS, Volume 1: edited by Gary E. McGuire

CHEMICAL VAPOR DEPOSITION FOR MICROELECTRONICS: by Arthur Sherman

CHEMICAL VAPOR DEPOSITION OF TUNGSTEN AND TUNGSTEN SILICIDES: by John E. J. Schmitz

CHEMISTRY OF SUPERCONDUCTOR MATERIALS: edited by Terrell A. Vanderah

CONTACTS TO SEMICONDUCTORS: edited by Leonard J. Brillson

DIAMOND CHEMICAL VAPOR DEPOSITION: by Huimin Liu and David S. Dandy

DIAMOND FILMS AND COATINGS: edited by Robert F. Davis

DIFFUSION PHENOMENA IN THIN FILMS AND MICROELECTRONIC MATERIALS: edited by Devendra Gupta and Paul S. Ho

ELECTROCHEMISTRY OF SEMICONDUCTORS AND ELECTRONICS: edited by John McHardy and Frank Ludwig

ELECTRODEPOSITION: by Jack W. Dini

HANDBOOK OF CARBON, GRAPHITE, DIAMONDS AND FULLERENES: by Hugh O. Pierson

HANDBOOK OF CHEMICAL VAPOR DEPOSITION, Second Edition: by Hugh O. Pierson

HANDBOOK OF COMPOUND SEMICONDUCTORS: edited by Paul H. Holloway and Gary E. McGuire

HANDBOOK OF CONTAMINATION CONTROL IN MICROELECTRONICS: edited by Donald L. Tolliver

HANDBOOK OF DEPOSITION TECHNOLOGIES FOR FILMS AND COATINGS, *Second Edition:* edited by Rointan F. Bunshah

HANDBOOK OF HARD COATINGS: edited by Rointan F. Bunshah

HANDBOOK OF ION BEAM PROCESSING TECHNOLOGY: edited by Jerome J. Cuomo, Stephen M. Rossnagel, and Harold R. Kaufman

HANDBOOK OF MAGNETO-OPTICAL DATA RECORDING: edited by Terry McDaniel and Randall H. Victora

HANDBOOK OF MULTILEVEL METALLIZATION FOR INTEGRATED CIRCUITS: edited by Syd R. Wilson, Clarence J. Tracy, and John L. Freeman, Jr.

HANDBOOK OF PLASMA PROCESSING TECHNOLOGY: edited by Stephen M. Rossnagel, Jerome J. Cuomo, and William D. Westwood

HANDBOOK OF POLYMER COATINGS FOR ELECTRONICS, *Second Edition:* by James Licari and Laura A. Hughes

HANDBOOK OF REFRACTORY CARBIDES AND NITRIDES: by Hugh O. Pierson

HANDBOOK OF SEMICONDUCTOR SILICON TECHNOLOGY: edited by William C. O'Mara, Robert B. Herring, and Lee P. Hunt

HANDBOOK OF SEMICONDUCTOR WAFER CLEANING TECHNOLOGY: edited by Werner Kern

HANDBOOK OF SPUTTER DEPOSITION TECHNOLOGY: by Kiyotaka Wasa and Shigeru Hayakawa

HANDBOOK OF THIN FILM DEPOSITION PROCESSES AND TECHNIQUES, *Second Edition:* edited by Krishna Seshan

HANDBOOK OF VACUUM ARC SCIENCE AND TECHNOLOGY: edited by Raymond L. Boxman, Philip J. Martin, and David M. Sanders

HANDBOOK OF VLSI MICROLITHOGRAPHY, *Second Edition:* edited by John N. Helbert

HIGH DENSITY PLASMA SOURCES: edited by Oleg A. Popov

HYBRID MICROCIRCUIT TECHNOLOGY HANDBOOK, *Second Edition:* by James J. Licari and Leonard R. Enlow

IONIZED-CLUSTER BEAM DEPOSITION AND EPITAXY: by Toshinori Takagi

MOLECULAR BEAM EPITAXY: edited by Robin F. C. Farrow

NANOSTRUCTURED MATERIALS: edited by Carl. C. Koch

SEMICONDUCTOR MATERIALS AND PROCESS TECHNOLOGY HANDBOOK: edited by Gary E. McGuire

ULTRA-FINE PARTICLES: edited by Chikara Hayashi, R. Ueda and A. Tasaki

WIDE BANDGAP SEMICONDUCTORS: edited by Stephen J. Pearton

Related Titles

ADVANCED CERAMIC PROCESSING AND TECHNOLOGY, Volume 1: edited by Jon G. P. Binner

CEMENTED TUNGSTEN CARBIDES: by Gopal S. Upadhyaya

CERAMIC CUTTING TOOLS: edited by E. Dow Whitney

CERAMIC FILMS AND COATINGS: edited by John B. Wachtman and Richard A. Haber

CERAMIC TECHNOLOGY AND PROCESSING: by Alan G. King

CORROSION OF GLASS, CERAMICS AND CERAMIC SUPERCONDUCTORS: edited by David E. Clark and Bruce K. Zoitos

FIBER REINFORCED CERAMIC COMPOSITES: edited by K. S. Mazdiyasni

FRICTION AND WEAR TRANSITIONS OF MATERIALS: by Peter J. Blau

HANDBOOK OF CERAMIC GRINDING AND POLISHING: edited by Ioan D. Marinescu, Hans K. Tonshoff, and Ichiro Inasaki

HANDBOOK OF HYDROTHERMAL TECHNOLOGY: edited by K. Byrappa and Masahiro Yoshimura

HANDBOOK OF INDUSTRIAL REFRACTORIES TECHNOLOGY: by Stephen C. Carniglia and Gordon L. Barna

MECHANICAL ALLOYING FOR FABRICATION OF ADVANCED ENGINEERING MATERIALS: by M. Sherif El-Eskandarany

SHOCK WAVES FOR INDUSTRIAL APPLICATIONS: edited by Lawrence E. Murr

SOL-GEL TECHNOLOGY FOR THIN FILMS, FIBERS, PREFORMS, ELECTRONICS AND SPECIALTY SHAPES: edited by Lisa C. Klein

SOL-GEL SILICA: by Larry L. Hench

SPECIAL MELTING AND PROCESSING TECHNOLOGIES: edited by G. K. Bhat

SUPERCRITICAL FLUID CLEANING: edited by John McHardy and Samuel P. Sawan

IN MEMORIAM

Alan G. King
1924–2000

Alan G. King, a Member Emeritus of the American Ceramic Society, passed away on October 14, 2000. The manuscript for this book was nearing completion at the time of his death, and Anthony King, his son, finished assembling the pages as a testament to Alan's illustrious career.

Mr. King served in the Army during World War II. After the war, he attended the University of Utah, obtaining a Bachelor's degree in Geology and a Master's in Mineralogy.

In 1985, Alan received the first Technical Achievement Award as chairperson of the Advanced Ceramics Task Force which provided focused direction in the field of advanced ceramics to Ferro Corporation. He retired from there in 1991 as a group leader of the research and development division. His career produced almost a dozen patents in the field of ceramic cutting tools.

His long and productive association with the American Ceramic Society was appreciated. In 1975, Alan became a Fellow of the Society in recognition of his many notable contributions to the ceramics arts and industry; in 1991, he became a Member Emeritus for more than thirty-five years of dedicated service.

Alan maintained a laboratory in his home's basement and wrote technical bulletins and manuals as well as the manuscript for this book. He previously authored the book, *Ceramics in Machining Processes,* published in 1966 by Academic Press.

Contents

Foreword

This book is arranged in chapters that parallel the ceramic processing and analytical procedures. Up front are discussions on experimental design and laboratory safety. Following this sequentially, are the steps for: mixing, milling, slip preparation, mixing coarse-grained materials, forming, green machining, firing, and machining. The final presentation is the effect of pro-cessing on properties and property measurements. It is difficult to keep up with instrumentation as this changes so fast. One will find parts of this book a little out-of-date, by a couple of years. Fortunately, ceramics is a mature technology with the essentials moving slowly.

Each step in the manufacturing process has attendant problems associated with it. Along with these problems are measures to reduce them. Some typical problems are particle segregation, agglomeration, contamination, pressure gradients, adherence to tooling, and temperature gradients during drying and firing.

Emphasis is on the practicality of doing these operations in the ceramic laboratory; this limits the batch size. A one-to-one correlation bet-ween laboratory experiments and production scale experiments is tenuous at best. These differences occur naturally because of differences in scale and differences in the equipment used. Heat transfer through a small body is faster than through a large one. Shear during mixing is different in a bench

Hobart than it is in a large Eirich. Pressure gradients in a laboratory coupon during pressing are different from those in a 18" x 24" x 1.5" slab. In spite of these differences, the same principles and techniques apply, with due recognition to scale.

Once the laboratory work has been completed to a reasonable conclusion, there is the serious problem of technology transfer to a manufacturing plant and sequentially to the customer. The overwhelming problem with technology transfer is the allocation of human and capital resources. Because of risks with anything new, management usually starts small to test the market. Customers are often restrained by similar considerations and will not readily accept, or even test, a new product. For example, a new glass pigmentation system was developed that showed promise. In order for a customer to test the new system, he had to risk an entire glass tank batch, with the subsequent clean out, refractory failure, production lost, and compromis-ing good relations with his customers. He declined to do the test. A substantial risk reduces the chances of running a test. However, this is another subject and not within the scope of this book, but perhaps deserving of another volume.

Each section has practical hands-on suggestions on performing and sometimes avoiding certain tasks. There are many drawings and photographs that illustrate both the equipment and the accompanying procedures; the author has experience with the illustrated equipment. This does not preclude that a better choice might be made, especially since instrumentation is developing so rapidly. The intent of this book is to bring to the reader information that is not, or is sparsely available, to people working in our industry.

Laboratory skills are gained by hands-on experience. The intent of this book is to accelerate this process.

Preface

Not long ago, I met a scientist who earned his Ph.D. degree, from a well-recognized university, in materials science (ceramics). During the course of conversation, this scientist mentioned to me that when he started to work in industry he had to acquire additional information to do the job. This declaration should shock many people, as it did me. After six to eight years of advanced education, it appeared that this scientist had not acquired enough information to do his job. Sadly, this is not an isolated case. Many companies, in a variety of industries, have encountered this problem. It will be constructive to try to understand this gap in training, specifically in ceramics. There are three main reasons that can account for this situation.

Firstly, ceramics is considered a technology and not a science. In that most properties of ceramic materials are predictable only from experience and not from theory. Kingery's book, An Introduction to Ceramics, marked a turning point in ceramic education with an emphasis on the underlying scientific principles. This text, along with subsequent texts by other authors with a similar mind set, is widely used throughout academia. This trend toward teaching ceramics science is indisputably laudable, but in some respects may have introduced a problem of exclusion. The interval within which a student can be subjected to bondage has both practical and humane limits. A student cannot possibly learn all the essentials needed even within the confines of materials science, let alone the laboratory crafts. So, there has to be a knowledge gap.

Secondly, scientific and engineering journals have severe restrictions on page space. The volume of papers submitted for publication exceeds the space available for publication; this necessitates exclusion of some material. Anyone who has written a paper has experienced this compression. Editors expurgate or excise descriptions of procedures to the extent that one cannot duplicate the work since it lacks some necessary information. This phenomenon creates a knowledge gap.

Thirdly, industrial research is driven by economics; this is a demanding taskmaster. The ceramic engineer is always balancing on a tight rope with regard to the following: cost verses performance, engineering verses creativity, empiricism verses engineering, and time verses resolution of the problem. The complexity of essential needs mitigates against a formalized approach towards structured education. Competition drives performance up and prices down, with these often in conflict. In industry, there will be a wide variety of problems each of which has its own demands on technique that cannot possibly be comprehensively included in a curricula. This further induces a knowledge gap.

Anyone entering the ceramics field is confronted with the above problems. Thus, they need to learn a great deal about laboratory procedures that were not taught in school and are not available in the technical literature. Additionally, technicians without a formal education in ceramics are even worse off, as they have to start with the basics. Herein lies the problem that raises the need for a solution.

This book is a compilation of laboratory experiences in the sequential phases of ceramic processing and analysis. While not in the style of a hand-book, in a real sense, it is a guide and manual to some practical aspects excluded from other available sources. This exclusion created a problem where the author, by necessity, had to draw largely from his own experience. This experience-oriented work leaves us with some under-represented areas and other over-represented areas. In the absence of an alternative, this book is written with the confidence that it is sufficiently comprehensive to be useful.

Is this a solution to the information gap in the curricula and technical literature? Of course not, but this is a start. With good fortune, this book will go into a second edition, in which case an expansion is possible to additional volumes on the neglected subjects, provided others are inspired to write them.

Acknowledgments

Since this book is about crafts and since little of this information is present in the literature, there are very few references that can be cited. Craft information is casual for the most part and the source is not easily recalled. Most references are through personal communication when such communication is remembered; I apologize to those whom I have forgotten. I thank the University of Utah (my Alma mater), the U.S. Geological Survey, Norton Company, Zircoa Corporation, Ferro Corporation, and the numerous people therein for their support and the opportunity to learn the ceramic crafts that are conveyed in this book. I would like to thank the following people at Ferro: Dave Gnizak and Alex McMaster helped extensively with the photomicrographs. Debra Oberlander and Carol Bican contributed immensely to the literature. San Keswani, Prasad Shertukde, and Tom Vlack taught me valuable computer skills. Dave Harrison was generous in his support. Dave Marchant and Mike Dowell were especially helpful with suggestions on the manuscript. The people in the Advanced Ceramic Group and ceramists in the plants also widely contributed to the information in this text.

1

Introduction

1.0 INTRODUCTION

This chapter addresses five subjects. All these subjects are non technical. They include such topics as objectives, managing data, reverse engineering, information sources, and acknowledgments.

2.0 OBJECTIVES

There is a gap in the technical literature concerning the crafts of ceramic engineering as there is very little information that is available regarding this topic. This topic deals with how to engineer crafts in a ceramic laboratory. Each generation of ceramic engineers learns from directly engineering the craft and from knowledge that they derive from their colleagues. In this book, I will try to bridge the gap in the technical literature by attempting to more efficiently describe these laboratory crafts.

John T. Jones and M. F. Bernard wrote a book entitled *Ceramic Industrial Processing and Testing.*[1] This book has two parts: part one is about ceramic industrial equipment and processing; part two is about

1

measurements and calculations. This current book supplements part two and broadens part one to include laboratory equipment and processing. A book by James Reed entitled *Introduction to the Principles of Ceramic Processing*[2] is more theoretical in content than the current book. With these three books, the subject of processing and procedures will be appropriately addressed.

Scope

The scope of this book covers a broad range of ceramic technical crafts, with some notable exclusions. Since comprehensive literature on ceramic lab crafts is unavailable, I had to largely deduce this knowledge from my own experience. While knowledge of these crafts is also applicable to electronic ceramics, this discussion does not directly address electronic ceramics, glass, or composites. Addressing these aforementioned materials is beyond the scope of this book.

Dealing with ceramic crafts in a manufacturing plant can be quite different than in a laboratory. Processing techniques in a manufacturing plant are different principally because of the volume of material involved and the urgency in getting an order out the door. However, there is an overlap, especially in analytical techniques and in instrumentation. In regards to processing, this book addresses laboratory procedures.

The scope includes information on preparing the material, forming the material, and firing the material. Also discussed are laboratory analytical procedures, testing procedures, and the properties of the ceramic that results from this processing procedure.

Various types of instruments and processing equipment are referred to in this book. To avoid any type of endorsement, it is not common practice to specify equipment by their brand name. As such, technical literature quite properly avoids identifying equipment by brand name. However, this book is not considered technical literature in the normal sense; it is about engineering crafts in a ceramics laboratory. Since equipment is a central part of this book, the author would be remiss if he did not include equipment brand names. Mentioning the brand name can implicate two problems. First, with progress being so rapid it is inevitable that some information will be out-of-date, and second, my experience will not be all inclusive. As a result, I decided to draw from my limited and sometimes outdated experience regarding the choices of the instruments and process equipment in this book. Please consider these limitations and do not be hesitant to take a fresh look when considering a new instrument or process equipment.

The general objective is to present in the technical literature an assembly of information on the crafts that a ceramic engineer or technician can use in the laboratory. This information will be helpful since it is not found any where else.

3.0 MANAGING DATA

The number of independent variables and anticipated interactions in a system has a lot to do with how an experiment is designed. We will discuss three such contingencies.

Models

Models are mathematical expressions derived from scientific or engineering principles. The expressions predict how the system is expected to function. The next step is to do the experiments to see if the model fits the data, or vice-versa. This is of course the scientific method. Provided one conducts the experiment scientifically, the model is confirmed when there is coincidence between the expected and the predicted results. The beauty of science is that it is self correcting. Anyone with the appropriate skills can run the same experiment and get the same results. Another beauty of science is that it is a discipline simplifying our concept of the physical world. Reducing a system to a working, mathematical expression imparts fundamental understanding and this is beautiful.

In important cases, the model becomes a *theory* when one experimentally verifies a mathematically derived concept. With very important fundamental cases, the model becomes a *law* when the experimental results always work. Results in the ceramics lab are usually of a more humble stature such as a relationship and correlation between a cause and an effect in a particular system. One can work with this correlation as it describes how the system behaves, and this is often just what one seeks.

Statistical Experiment Design

When the independent variables are mostly known, or can be anticipated, statistical experimental design is sometimes the preferred method. Factorial designs are particularly powerful. For example, consider a small sintering experiment shown in Table 1.1.

Table 1.1: Sintering Experiment

		Sintering Temperature		
		Low	Medium	High
Molding Pressure	Low	2	2	2
	Medium	2	2	2
	High	2	2	2

Table 1.1 depicts two independent variables: sintering temperature and molding pressure. One can expect to observe curvature as each of the independent variables is at three levels. Remember, two points define a straight line. To observe a curvature, there has to be as a minimum of three points: low, medium, and high.

Let us postulate that the dependent variable is sintered density. There are two samples in each condition; enabling us to find an error variance. The total number of samples in the experiment is 18. One can compare the effects of temperature and pressure on density to the error variance with an F ratio. One can find the latter value in a book to determine the probability that the effects are real or just due to chance. This provides a much more secure basis for making decisions.

Factorial design also produces an interaction variance that indicates whether the two independent variables have an effect on each other. This is called the curvature. In this example, one would expect an interaction. Often, an interaction is the most important result from the experiment and this is the way to obtain it.

Factorial designs are fine when there are only a few independent, known variables. When there are four or five such variables, the experimentation becomes lengthy. Partial factorials can help to reduce the bulk, but some interactions are lost.

There are many factorial experimental designs; some of these designs are multidimensional. Selecting the best design for an experiment requires some expertise. To achieve this end, one can either consult an experimental statistician or be further trained in this area.

Another method of statistical design is *Self Determining Optimization* (SDO). This is a method that leads one in a stepwise fashion to the best solution. In SDO, one performs a sequence of experiments that lead in the direction of the desired result. SDO is analogous to the game called *Blind Man's Bluff* that also leads you in the right direction. Both these methods provide the person with the right answer rather than an understanding of the system. Since the right answer is important, one benefits from applying these methods. SDO is especially useful when there are a lot of independent variables that render factorial methods impractical.

Other useful statistical measures are standard deviation, coefficient of variation, and the correlation coefficient. When two variables plot as a line, these statistics indicate how good a fit there is in the data.

Table 1.2 and Figure 1.1 depict three sets of data in a decreasing certainty of correlation; this is measured by the correlation coefficient.

Table 1.2: Curve and Correlation Coefficient

Curve	Correlation Coefficient
Good Fit	0.997
Poor Fit	0.938
Bad Fit	0.58

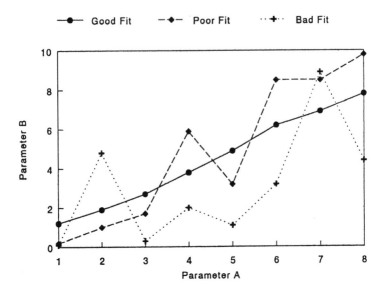

Figure 1.1: Linear Curve Fitting

Figure 1.1 represents some nonsense data to make a point. It depicts a visual comparison between good, poor, and bad data in correlation to a straight line. In this example, the visual insecurity in the data is compared with the statistic, *Correlation Coefficient.* The author's preference is to accept a good fit as a mathematical relationship and the poor fit as a trend, but would balk at data scattered widely as in the bad fit. It is not difficult to calculate the linear equation for the good fit line from the slope and intercept. For the above data the equation is:

$$A = -0.01 + 0.986B \tag{1.1}$$

Equation 1.1 can be used to calculate other values of A and B and to extrapolate accordingly, provided one does not go too far.

Standard deviation (SD) is a statistic that measures the scatter in a set of data. Coefficient of variation (CV) is the standard deviation divided by the mean (average). Coefficient of variation is used to compare two standard deviations when the mean is different between the two sets of data. Consider the nonsense data in Table 1.3.

Table 1.3: Data with Standard Deviation and Coefficient of Variation

	Set A	Set B
	50	42
	48	58
	54	42
	47	58
	48	56
	57	54
	46	56
	54	50
Average (mean)	50.5	50.75
Standard Deviation	4	7.17
Coefficient of Variation	0.079	0.141

The mean values are almost identical, but the standard deviations and coefficients of variation are different. Set B has more scatter than set A, as seen in Figure 1.2.

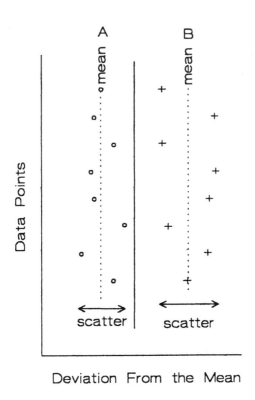

Figure 1.2: Scattering in Data

Figure 1.2 is a visual comparison between two sets of data with one having more scatter than the other.

It is always a good idea to look directly at the raw data to help to obtain a feel for the statistical significance. In Figure 1.2, the scatter from off the mean is shown, with set B having more scatter than set A. Visually the scatter in the figure imparts meaning to the statistics. There is another reason for looking at the data. An inconsistent value is obvious and will need an explanation. For instance, a set of identical numbers from the modulus of rupture (MOR) test does not mean that the strengths were perfectly identical; this is very close to impossible. Rather it means that there is an instrumental error. It is more common to have one or two numbers that do not fit with the others. For example, if set B had a value of 85, it would be suspect and most statisticians would eliminate it from the set. One can assume that something went wrong and the value is not representative.

Are the results reasonable? Often the experimenter will have a good idea of just what to expect from a test. When the result is good compared with expectations, then either a discovery or a mistake was made. One would have to verify this.

Let us now address the coefficient of variation. Let us assume that we have a third Set C (set not shown) with data that is numerically the same as set A, except that the SD and mean is smaller by a factor of 10. This produces the following statistical result:

	Set A	Set C
Standard deviation	4.0	0.4
Mean	50.5	5.05
Coefficient of Variation	0.079	0.079

The two sets of numbers have different standard deviations and means, by a factor of 10. However, the coefficients of variation are identical, meaning that sets A and C have the same scatter in proportion to their mean values. It is not useful to compare standard deviations between two sets of numbers when the mean values are greatly different. In such a case, use the coefficient of variation instead.

Another way to present data in a meaningful way is to draw contour maps. Contours are drawn through the data points to delineate the locus of equal values. Take for example the data in Figure 1.3. This figure is hypothetical and does not represent a real system. Obviously, there is a wealth of information in a contour map of this sort. There is a maximum, a slowly sloping plateau off to the left, a ridge-like bulge, and then a precipitous drop off on the far left of the map. Parameters A and B are interacting in an active way. An especially important advantage is the almost instantaneous recognition of the system's behavior. This may not occur by merely looking at a page full of numbers.

When drawing the contours over a field of numbers, inconsistencies are to be expected as data has random variability. Surfaces usually behave in a consistent pattern. Contours are not independent of each other but are interactive, with the shape of the surface evolving as it is drawn. When a few numbers are not in the right place, it is of little consequence. For example, the figure "93" is on the wrong side of the "95" contour. To make an abrupt change in the contour would compromise the consistency of the pattern and would probably be a false description of the surface shape.

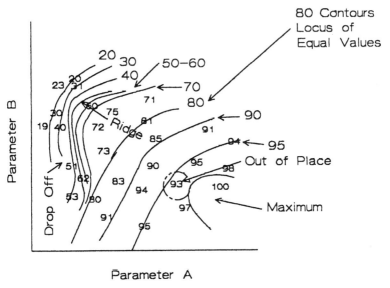

Figure 1.3: Contour Maps

As some computations are lengthy, computers are especially useful for statistical analysis. ECHIP is user-friendly software; it also has a user-friendly manual. STATISTICA is more powerful software, but it is complex to learn and is expensive. CORELDRAW, a graphics software, is one of several good programs for graphics application.

There is a cautionary word about time spent on graphics software application. Data reduction and presentation should not be time consuming and should not be a substitute for experimental time in the laboratory. Simple sketches are often good enough. Some of the most descriptive figures I have seen were those drawn by professor Newnham at Penn State; these were simple drawings. Anatomists frequently use drawings to selectively emphasize particular features. As such, a class in drawing techniques can be a substitute for a class in computer graphics. Attaining such drawing skills could be time well spent.

A computer is not a surrogate brain. Similarly, a copy machine is not a surrogate pen. There isn't always a need for long convoluted reports with gorgeous graphics when one only needs a contour map. A contour map can be sketched in a fraction of the time it takes to produce it with a graphics program; moreover, it conveys the same amount of information.

Intuitive Experiments

It is axiomatic that if nothing new is tried that any new discovery is precluded. Creativity in the physical sciences seems to involve inductive processes. A creative individual tries new things with the process often being intuitive. A memorized fact, an observation, or information from a different field can often spark an intuitive experiment.

Despite the fact that the subject of creativity has been extensively addressed, this topic still leaves the author without an understanding of the creative process. However, the following factors contribute toward the art of creativity:
- latitude in time and resources for individual research,
- an environment appreciative of invention,
- an environment where invention is required,
- resources for commercialization particularly in an industrial lab,
- an inquisitive mind, and
- stimulating colleagues.

Another impetus for intuitive creativity is the recognition of a problem with subsequent probing for an answer. The key factor involves understanding the problem. For example, consider the problem with forming recrystallized silicon carbide: the ceramic has little or no firing shrinkage; however, making the green density can be the key to obtaining a high-fired density. Then, start thinking about what processes might increase packing density. Some of these processes include gap sizing, mulling the grain to round the corners, vibrating the mix as it is slip cast, jolting the mix as it is slip cast, vibratory pressing, surface chemistry for better dispersion, and rolling the slip with coarse grain additions. These processes will be discussed in Chapter IV. In the case of the author's experience, there is adiabatic injection molding where the amount of water can be reduced in the mix; this is described in chapter 4.

The author still does not have a good understanding of how to stimulate the creative process. However, the following factors can squelch the creative process: rigid routine, negative acceptance, lack of resources, and constant failure to commercialize the creation.

4.0 REVERSE ENGINEERING

Analytical Methods

Reverse engineering is a catch-up strategy where a competitor's product is legally acquired and analyzed. For ceramics, these analyses include dimensions, shape, appearance, chemical composition, phase composition, mechanical or electronic properties, and microstructure.

Armed with this information, the problem becomes one of duplicating the part. Sometimes, there are marks on the part that provide clues about how it was made. For example, a satellite sphere is usually die pressed, but if the diameter is small it is probably rotary pressed. Additionally, if there is the remnant of a gate and it is fine grained, it is injection molded. Anyway, knowledge of ceramic engineering is applied to duplicating the part.

There is a business problem here. A duplicated product will by definition be "Me Too" and it is difficult to successfully enter a market where the competition is already entrenched. The top two, or in some cases the top three, make money and the rest just drag along.

A superior strategy is to craft an advantage that is perceived by the customer. Price is the predominant motivator, provided that the product quality and delivery are competitive.

To make a profit from a reverse engineered product, analysis of the product is only the first step. The important step is to find a way to make it better, and even this does not guarantee sales.

Let us consider the following example. A salesman called on a purchasing agent to sell him a reverse engineered product. The purchasing agent was initially very enthusiastic about the product. However, after the sales pitch, the purchasing agent leaned back, and with a smile, said, "Son, you are wasting your time. The salesman from the other company is godfather to my kid, and we have dinner at my house and play pinochle every Friday, not to mention, sharing holidays with the families." As such, there is little reason to call on this account under these circumstances as sales are locked in by personal attachments.

Ethics

Industrial espionage is a criminal act. It is illegal to swipe a competitor's product or to bribe one of his employees to impart sensitive information. To stay out of trouble, check out the legality ahead of time. There are legal and ethical ways to acquire a product by either buying it or having it given to you by the owner. When you acquire it under these circumstances, it is legally and ethically yours.

The other aspect of ethical problems is to protect sensitive information within your organization. This is more of an issue for industry or defense R&D organizations. Generally, industry tends to suppress information beyond that which is prudent. Defense R&D automatically classifies much of their information as secret. Professional scientists, engineers, or technicians have an obligation to interact sensibly with their peers in regard to accessing and sharing information. Secrets should have a time restriction. A company should have the choice to commercialize, patent, or release the information. Sometimes, I have a vision of three different laboratories clutching tightly to the same secret information, this is ridiculous!

5.0 INFORMATION SOURCES

The computer has been described as a prosthesis for the brain, in the sense that huge amounts of information are easily available. One should expect this trend to increase to where it becomes pervasive. In addition to computer networks, there are other sources of useful information.

On-line Literature Searches

Electronic searches can be helpful, but not everyone has attained complete satisfaction with them. There are two problems: the bulk of the citations have no relevance to the subject of the search, and primary citations are often missed. While there is little to lose with using such a search, do not regard them as totally reliable. The problem seems to be in the choice of the key words by which the literature is indexed. This seems to be catch-can with everyone doing their own thing.

Older abstracts

Examples of such abstracts are Ceramic Abstracts and Chemical Abstracts. These are much more reliable than on-line searches; however it is time consuming to wade through them. One should use these sources for important citations.

Technical literature

Scientific Journals are routinely read and are the depository of information that is sought after from the literature searches. The better journals are peer reviewed; guaranteeing that all of the information on crafting the experiment has been expurgated. However, the results section incorporates the best information that can be found.

Thomas and other registers

The Thomas register contains an amazing wealth of resources. All of the companies listed in the register are anxious to help with goods and services. These volumes are comprehensive and are a further testament to the advanced industrial capacity of the United States.

Colleagues

Individual and personal contacts are a prime source of information. There is much that is not proprietary and can be passed along. This works both ways in that it is a conversation not an interrogation.

REFERENCES

1. J. T. Jones and M. F. Benard, Ceramics-Industrial Processing and Testing. Ames, Iowa: The Ohio State University Press, 1972.
2. James S. Reed, Introduction to the Principles of Ceramic Processing. New York, NY: John Wiley and Sons, 1991.

2

Safety

Had I more time, this letter would be shorter.   *Voltaire*

1.0 INTRODUCTION

Safety in and around a ceramics laboratory is by far the most important topic this book will address. Though brief, this is an important chapter that is worth your while to read and remember.

Ceramic Laboratory Hazards

A few years ago, a survey was conducted that compared the hazard potential of various types of laboratories. I wonder which laboratory type a reader would consider being most dangerous: a chemical lab involving daily work with poisons and corrosive acids, an electronic lab involving daily work with high voltages, a heat treatment lab involving daily work with red-hot metals, or a ceramics lab. Interestingly, a ceramics lab was the most dangerous of all labs surveyed. The ceramics lab contains work done on a daily basis at all the non-ceramics labs. Additionally, a ceramics lab is equipped with heavy, high-speed machinery, presses with high tonnages, and gas kilns that may explode.

Ways to Avoid Accidents

One should constantly think about safe working conditions in the lab. Some of these safe working conditions can be organized into safety inspections with frequent follow up. These inspections create a check list of problems that must be corrected immediately. In the author's lab, there was the following rule: No one has to work at a task that he/she considers to be unsafe, and there should not be any criticism to the individual involved. Before the individual can go on with the task, the problem in question has to be addressed and corrected to the individual's satisfaction. This was a good rule that made all accidents unconscionable.

A good book to read and constantly reference is <u>Prudent Practices in the Laboratory</u>, National Academy Press, Washington, D.C. The book contains the following nine chapters:
1. The Culture of Laboratory Safety,
2. Prudent Planning of Experiments,
3. Evaluating Hazards and Assessing Risks in the Laboratory,
4. Management of Chemicals,
5. Working with Chemicals,
6. Working with Laboratory Equipment,
7. Disposal of Waste,
8. Laboratory Facilities, and
9. Governmental Regulation of Laboratories.

Codes

Many crafts have codes that regulate the way the work is to be executed. Codes dictate the standards in each trade and are the best source of information for each trade. Often, these codes are legal requirements. If these codes are not observed, there could be resulting charges and fines. The supervisor's job is to diplomatically make sure that the appropriate codes are being followed to ensure safety for everybody.

Safety Check List

The safety check list is given below. Observe the following safety rules when working in the ceramics laboratory:
- organize periodic critical safety inspections;
- have both internal and external inspections;
- become an advocate on safety precautions;
- adopt a rule that no one has to work in a situation they consider unsafe;
- observe the precautions in the book, *Prudent Practices in the Laboratory*; and
- follow the required codes.

3

Milling and Equipment

1.0 PURPOSE OF MILLING AND MATERIALS

This chapter includes information on equipment and procedures for milling ceramic material. Milling produces a particular particle size distribution and deagglomeration of fine powders. Physical processes include impact, shear between two surfaces, and crushing by a normal force between two hard surfaces. When a solid is fractured, energy is given off as heat from fracture, friction in the equipment, and energy necessary to create additional surface area. It is the energy from creating additional surface area that does the work sought.

There are two broad types of ceramic raw materials that require milling. These are classified as lumpy and powdered ceramics.

Lumps result from mining, fusion, and sintering. These are usually premilled by the supplier and are available in various screen sizes. Depending on your requirements, these may require further milling in the lab. Mined materials include talc, shale (clays), bauxite, and quartz. Fused materials include fused alumina, magnesia, mullite, and zirconia.

Related to fusion is the Achesion furnace for making SiC. A pile of mix: sand, partially reacted material from a previous run, and coke is reacted by resistance heating at a very high temperature to form a "pig" of SiC that has to be crushed. Sintered raw materials include sintered clays, tabular alumina, and polycrystalline grogs (coarse granules) made from alumina, mullite, and zirconia. One way of making Si_3N_4 is to heat silicon metal in a nitrogen atmosphere. Carbo-thermal reactions at a high temperature are a common way to make a variety of carbides.

Powdered raw materials usually come from chemical processes where the material in solution chemically reacts to form particles in suspension. An important class is Bayer alumina that starts as bauxite. The Hall Cell fuses the bauxite with the cryolite flux forming a sodium aluminate that is then leached with caustic and washed to produce a moderately pure alumina powder. The powder is agglomerated and then milled for ceramic formulations. Several varieties are available according to crystallite size and purity.

In other processes, the material is dissolved and is then precipitated chemically as a powder. High purity ceramic powders are sometimes made from an organic precursor decomposed to make the ceramic powder. Included in this class are some alumina and zirconias with or without the stabilizer.

Much of this information is about ball mills used for processing ceramic slips. Emphasis is on fine-grained slips, with limited information on equipment for coarse materials.

Some materials are more difficult to mill than others. Generally, the order of difficulty from the most difficult to the least difficult is dense-fused materials, sintered materials, and precipitated powders. Although one might not expect this, glasses are very difficult to mill to micrometer sizes, but they are easy to crush to granules.

2.0 DRY MILLING

Production milling is sometimes done dry as this avoids a separate drying step. Dry milling also avoids the formation of hard agglomerates as there is no liquid present. Dry milled ceramics are usually used in pressing operations to make a shape and to consolidate the particles. Crushing and milling are sequential processes for particle size reduction. They will be discussed separately.

Crushing

Crushing helps reduce the particle size of hard materials to about 80 mesh using Tyler Sieves. After achieving this size reduction, other finer reduction techniques can be used in the lab. Two types of crushers are most commonly used: jaw crushers and roll crushers.

Jaw crushers

Jaw crushers have two hardened steel jaws, a stationary and a moving jaw. The moving jaw reciprocates in and out while exerting a crushing force on the granules. The cavity between the two jaws is tapered so that the finer particles drop down into the taper where they are then crushed to an even finer size. An adjustment on the width of the gap enables a jaw crusher to reduce the size of particles to about 10 mesh. Embedded particles on the jaw surfaces will contaminate subsequent batches. One can reduce contamination by running part of the new batch through the crusher and subsequently discarding it.

Roll Crushers

Roll crushers have two counter rotating steel rolls that are about 6 inches in diameter with an adjustable gap through which the material is crushed. These crushers can crush materials to about 80 mesh. Particles that press onto the roll surfaces will contaminate future batches. Wire brushing the roll surfaces reduces contamination, but the only way to completely clean the roll surfaces is to have them machined though this is not at all practical. Roll crushers are dangerous as loose clothing such as sleeves, neckties, necklaces, or gloves can be caught between the rolls. One should avoid wearing or using anything that can be caught between the rolls.

Milling

Ball milling

Usually dry milling is done in a ball mill with a milling media. Ball mills and media are discussed later in this chapter. Portland cement is dry milled commercially using steel balls for efficiency. An Alumina media about three inches in diameter is used for milling white cement where the color has to be controlled. Hard ferrites are also dry milled in air swept ball mills. Bayer aluminas are dry milled by the producer.

In the lab, dry milling is not as common since the cost of drying is not a factor. However, when the product, which is originally wet milled in the lab, is transferred to production where it is dry milled, the properties of the powders will be different and will cause start-up problems. Dry milling is not effective in small diameter mills as it lacks enough energy to fracture the particles. Mills with about a 12-inch diameter are suitable. Higher energy mills such as planetary and stirred mills are also available.

A serious problem with dry milling is that the powder will cake onto the sides of the mill and will not receive further size reduction. Scraping the mill down periodically helps to reduce this caking problem. Another way to reduce the caking problem is to add a surface active agent. It is believed that Bayer alumina dry milling involves the use of ethylene glycol. Bone-dry powders have less of a caking tendency than a powder exposed to humidity. Generally, powders can be dried and immediately put into the mill while still hot.

Jet Mills

Jet mills have two opposed jets of air that collide. These air jets also contain the ceramic particles. Often these mills are connected with an air classifier and a cyclone for recovering the fine particles. This set up is much more capital intensive than ball mills. It is also more difficult to clean between batches. Jet mills require a lot of high pressure air (80 psi). As the particles are in free flight, there is little contamination. However, a white powder will turn grey after milling due to the polymeric mill lining.

3.0 WET MILLING

Wet milling is more frequently applied in the laboratory than dry milling. Wet milling is usually used to make a coarse-grained slip or a fine-grained slip.

A slip consists of a liquid vehicle, usually water, and suspended ceramic particles. Fine-grained particles are held in suspension by dispersants and other surface active chemicals. Settling is basically by Stoke's law where larger and denser particles will settle out faster in a low viscosity liquid. This will be discussed further in a later section. Wet milling reduces the particle size for fine grained slips and disperses the agglomerates in both fine and coarse grained slips.

4.0 EQUIPMENT

Jar mills

These mills are cylinders with a capacity of about 0.5 liters to 2.0 liters. Many labs have larger equipment depending on their specific needs. Generally, slips in the lab are made up in smaller quantities. Jar mills are generally rotated on mill racks rather than with an individual drive.

Porcelain Mills

These mills are made from an alumina fortified porcelain. Porcelain has dispersed alumina particles in a glassy matrix. Mills and lids are slip cast, dried, and fired. The porcelain body has a composition of about 60-70% alumina with the remainder being a glass mostly silica with some alkaline earths, usually magnesia and sometimes some baria. Bayer processed aluminas are used in the formulation. These contain about 0.4%

Na$_2$O in the glassy phase and about 2-3% Na$_2$O that provides some fluxing and softening of the glass. Microstructure shows that there is about 10% residual porosity in the body.

Porcelain bodies are not as wear resistant as 85-90% aluminas by a factor of about four. However, most of the batch contamination during milling is from media wear rather than from the mill body. Porcelain mills are tolerable when the contamination by alumina and silica is acceptable.

Figure 3.1: Typical Alumina Fortified Jar Mill (courtesy of E.R. Advanced Ceramics).

Figure 3.1 shows both porcelain and rubber jar exteriors. Porcelain has to be rotated on rubber rollers to obtain enough friction. Rubber exteriors can be rotated on either steel or rubber rollers.

Hand tightening the closure is all that is required if the mill lip, gasket, and lid are kept clean. Most mill leakages occur due to failure to keep the sealing surfaces clean. Unclean surfaces can lead to over tightening and a deformation of the cross bar. Slips containing a binder dry harder though the binder can be scraped away. Porcelain is brittle and can break if it is dropped or falls off the mill rack.

High Alumina Jar Mills

These mills have an alumina content of 85-90% with the remaining being silica, alkaline earths, magnesia, and sometimes baria. Calcia is also, but rarely, used in some compositions. Bayer alumina is the principal source of alumina. It contains about 0.4% Na_2O, which is appreciable when dissolved in the glassy bond. The microstructure has small (3-5 μm long) alpha alumina crystallites faceted in a glassy matrix. Dealers often do not distinguish between porcelain and high alumina, but there is a way to distinguish them. Porcelain is more translucent with a light grey cast, but high alumina is white and opaque. This difference is evident when the jars are held up to bright light. There is some residual porosity in the body. These pores are unusual in that they are large (10 μm) and are scattered throughout the structure in fully dense regions of crystallites and glass. These pores are probably caused by degassing the ceramic after the structure sinters to an impervious state; when gas can no longer escape and causes the body to bloat. Wear resistance of high alumina is dependent upon the alumina content, volume percent porosity, and crystallite size. Optimum wear resistance is approached with 90% alumina, less than 5% porosity, and a 3 μm alumina grain diameter. Crystallite size depends on the firing temperature, which can be reduced when the green density is high. The manufacturer may have photomicrographs for perusal, or a piece which can be polished and observed. Like porcelain, alumina is a brittle ceramic and can break if it is dropped.

Polyurethane Lined Mills

Steel or polymer mill shells can be fitted with a polyurethane liner inside the mill and on the lid. Polyurethane is tough and slow to wear and is a good choice for use in a lab mill because contamination is minimal. However, polyurethane is not resistant to most solvents as it will swell or dissolve. Some solvents destroy the mill liner, and this damage is irreversible. Therefore, polyurethane liners are usually limited to water slips. Polyurethane is, however, also documented as being resistant to a variety of solvents. Polyurethane lined mills are resistant to fracture if dropped.

Neoprene Lined Mills

Neoprene liners are available for steel mill jars and lids. Neoprene is resistant to some solvents and is better than polyurethane in this respect. It is a strong and tough elastomer. Unfortunately after the milling process, there may be a black residue of neoprene floating on the surface of the slip. This could be a problem depending on how sensitive the ceramic is to this contamination.

Steel Mills

Steel mills are very useful whenever contamination from the ceramic mills or media is intolerable. One example is milling SiC. Contamination from the steel can be leached out with an acid, while contamination from porcelain cannot. Steel is relatively inexpensive and of course being malleable, it will not break when dropped. A steel grinding media has a specific gravity of about 7.8 g/ml compared with high alumina at 3.5 g/ml. The higher density imparts more energy to milling resulting in shorter milling times to reach a given particle size distribution.

High Purity Ceramic Mills

These mills are very expensive, but commercial sources can be found. The problem being that it is very difficult to slip cast relatively large shapes from these fine sized slips. However, jars made from high purity alumina, yttria stabilized zirconia, or silicon nitride have been made and used. One example is the milling of silicon nitride (Si_3N_4) where steel milling is not preferred because the acid leach affects the ceramic powder. In this case, a silicon nitride mill and media are used.

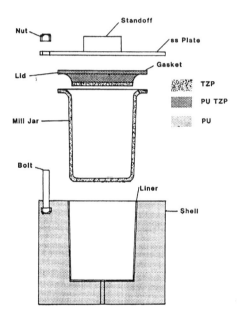

Figure 3.2, Ball Mill Assembly

0.1 m

Figure 3.2: Tetragonal Zirconia Polycrystal Experimental Mill

Figure 3.2 is a drawing of a zirconia high purity mill jar in a polymer encasement.[1] This mill is non contaminating and is without dead areas. This is an experimental design; it is included here because it addresses some problems with jar mills. The ceramic-yttria-stabilized zirconia is often referred to as Tetragonal Zirconia Polycrystal (TZP). Casting grade polyurethane was used for the encasement. The gasket is polyurethane filled with TZP powder mixed on a three roll mill. By casting the gasket in place on the mill jar, a perfect fit is achieved. Most gaskets are not this tight and some slip penetrates into the join and is therefore not milled. When this happens, agglomerates get into the slip lowering the ceramic strength. This design allows the jar to be lifted out of the encasement for a clean recovery of the slip. With most jar mills, the slip is poured out over the lip and the top of the mill body; this adds contamination to the batch. However, contamination does not occur here as the seal is dry and with the jar removed, the slip is not poured over the top of the mill body. Another design is shown in Figure 3.3.

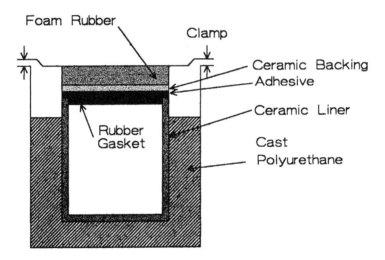

Figure 3.3: High Purity Vibratory Mill

There are several parts to the mill: a ceramic jar, a lid with a gasket, a polyurethane casting that holds the jar in place, a foam rubber piece that holds the lid in place, and the stainless steel can and lid. The jar is removable. The milling media in this application is 3 mm in diameter and is half filled with TZP. The mill is clamped to a vibrating base.[2] Milling is not nearly as intense as in a ball mill. In this case, the batch is not contaminated partially due to the design and partly due to the TZP media that is extraordinarily wear resistant. During vibration, the batch does not splash up to the gasket; eliminating another source of contamination. As was stated earlier, this design is not readily utilized due to the deagglomeration of soft particles. Incidentally, stainless pots and trays are a lot less expensive when bought from a restaurant equipment supply house or department store.

Dead Spots

Milling agitation is not uniform in a jar mill due to the shape of the interior cylinder that has corners at the lid end and at the base of the mill. The lid end can be especially troublesome as unmilled material can accumulate especially in the recess around the gasket. This is a dead spot that allows the unmilled or under-milled particles to enter the batch when the mill is dumped. It would be convenient if the procedure was adapted to interrupting the milling process to clean out the gasket area at least once. This would eliminate the accumulation of large agglomerates should the material be sensitive to agglomerates.

5.0 MILLING MEDIA

The jars hold the batch and the milling media along with the liquid that is usually water. It is the milling media that does the grinding. There are a substantial number of media choices.

Type of Media

Composition of the media includes the following: porcelain, high alumina, pure alumina, TZP, MgO stabilized zirconia, silicon nitride, silicon carbide, steel, modified fused zircon,[3] and a variety of mineral products such as flint, agate, or the material that is ground by itself (autogenous milling). Mineral products are cheap and can be surprisingly wear resistant. When processing fine ceramics, a general rule is to use the same composition of the media as that of the batch, if possible. For general lab applications, high alumina is perhaps the most commonly used media.

Size and Shape

Depending on which media is selected, the size ranges from 1 mm to 3 inches in diameter. The choice of size depends on the material being milled. A general rule is to use the smallest size that has sufficient energy to fracture the particles in the batch.

The best media shape, to obtain optimum grinding efficiency, is debatable in current literature. Most common shapes are spheres, satellite spheres, cylinders, and round-ended cylinders. Media with sharp edges should be prerun in the mill to round off the edges; otherwise, they will tend to chip and contaminate the batch. Stirred mills present a special problem as the media can pack and overload the drive. In such cases, spherical or satellite spheres are preferred.

Fill

Jar mills are filled about half full with grinding media. When the jar is over filled, there is not enough space for the media to tumble. When under filled, there is excessive media wear. Media that cascade with too much energy can chip or break. Sand mills have a vertical mill case and a vertical stirrer. These mills use small spheres such as quartz sand, alumina, or zirconia media. When the media is too dense, the weight of the media can prevent churning (the action that does the milling). However, horizontal stirred mills do not have this problem.

Milling Rate

The milling rate for a given batch formulation is affected by several factors such as the mill type, media size, and media specific gravity.

Mill type

Stirred mills are faster than jar mills. Large diameter jar mills are faster than small diameter mills as the media cascade drops over a greater height. Speed depends on the amount of energy in the media created by the agitation. However, media energy is limited when the media starts to fracture.

Media size

There are more points of contact per unit volume for smaller diameter media than for larger diameter media. This increases efficiency. Stirred mills impart additional energy, and they can get away with using smaller diameter media. Lots of stirred mills use media that are 2-3 mm in diameter. However, small diameter media are more expensive than larger diameter media as the cost curve is logarithmic to diameter size reduction. Due to cost concerns, newly designed stirred mills try to use a lesser quantity of media. However, it is important to consider the strength of the media. If the media is too weak, it will fracture and contaminate the batch. In this respect, TZP media is often the best choice, though expensive. High quality TZP is very strong and thus highly wear resistant. Since TZP lasts a long time and results in minimal contamination, it is sometimes the preferred media and should be regarded as an excellent investment.

Specific Gravity

The specific gravity of the media dictates the energy in the mill. High density mill balls have more energy because of their mass. Table 3.1 gives the specific gravity of commonly used materials.

Table 3.1: Specific Gravity of Commonly Used Materials

Material	Specific Gravity g/ml
WC/Co	13.0 - 15.0
Iron	7.8
TZP	± 6.2
Zircon	4.7
Alumina	3.98
Silicon Carbide/ Silicon Nitride	3.21/3.18
Porcelain	± 2.7
Flint	± 2.62

Media Wear

Wear tests can be done by placing a specific number of specimens (6) in a small jar mill with a 3 mm TZP media. These specimens are about the same size as the TZP media, but they vary in shape from spheres to cubes depending on what is available. The data that follows is based on two slips: A-16 sg alumina and 600 grit Silicon Carbide (SiC). After each run, the specimens were picked out with tweezers, washed, dried, and weighed. Each type of media in the test was identifiable by its color, shape, or size. As many as 8 types of grinding media can be run together and a fresh slip was used for each cycle. Table 3.2 summarizes the ceramic material and slips that will be shown in greater detail in Figures 3.4 to 3.9.

Table 3.2: Ceramic Material vs Slips Used

Figure	Ceramic Material	Slip
3.4	Aluminas	A-16
3.5	Zirconias	A-16
3.6	Silicon Carbide & Silicon Nitride	A-16
3.7	Aluminas	SiC
3.8	Zirconias	SiC
3.9	Silicon Carbide & Silicon Nitride	SiC

The experimental data is presented in Figures 3.4 to 3.9. Each of these figures will be discussed in relation to the different sources of materials and in relation to each other.

In Figure 3.4, four sources of alumina grinding media were tested with the A-16 sg slip. Sample one is 99.99% alumina, fully dense, and with a 2 μm grain size. Sample two is 85% alumina; a conventional grinding media. Sample three is a well crafted, high density media from Japan. Sample four is 99.9% alumina, fully dense, and with a 3 μm grain size.

In this slip, three of the materials were worn essentially at the same rate. Sample one was exceptional with half the wear rate compared with the other materials. This is attributable to its high purity, density, and fine grain size.

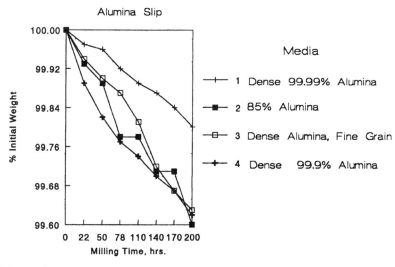

Figure 3.4: Wear-Aluminas, A-16 sg Slip

Figure 3.5 shows data for zirconia ceramics in A-16 sg slip. Samples one and two are TZP materials with 3 m/o Y_2O_3. Sample three is a magnesia stabilized zirconia with about 3.5 m/o MgO. This material is substantially less wear resistant than the TZP media. There is a large difference in wear rates for the TZP ceramics and since these materials are so expensive, it would be a good idea to run a test before a selection is made. Sample two is a much better buy than sample one; it is also less contaminating.

Zirconia does not necessarily outperform alumina as there are other factors such as microstructure that have to be considered. Wear results on both ceramics show that actual tests have to be done before wear characteristics can be predicted.

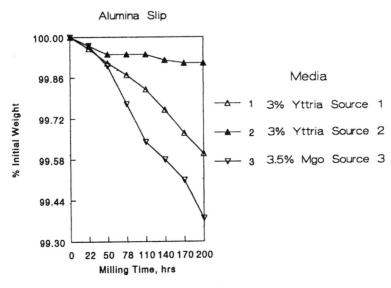

Figure 3.5: Wear-Zirconias, A-16 sg Slip

Figure 3.6 displays data with the Silicon Carbide (SiC) and Silicon Nitride (Si₃N₄) media in the alumina slip. Silicon Carbide does not show any detectable wear. The initial drop is probably due to a small chip that broke off the satellite sphere edge. Silicon nitride shows an initial, no-wear zone before it starts to wear at a slow rate. (The figures are at different scales, so it is necessary to read the values on the ordinate.) It seems that a tough skin on silicon nitride prevents initial wear. This seems real as there are four data points, each of which is the weight loss (none) of six samples.

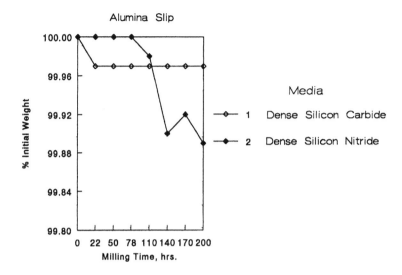

Figure 3.6: Wear-Silicon Carbide/Silicon Nitride, A-16 sg Slip

Figures 3.7 to 3.9 display data for the above ceramics but in a silicon carbide slip.

Figure 3.7 displays the data for alumina ceramics. Three of these ceramics behaved in a similar fashion, wearing at a higher rate in the SiC slip as expected. The other alumina, the 85% material, wears at a much higher rate than the others. This is comparable to the alumina slip and suggests a change in the wear mechanism..

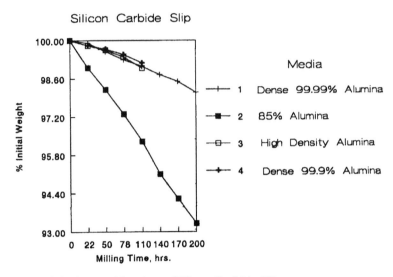

Figure 3.7: Wear-Aluminas, Silicon Carbide Slip

Zirconia media wear results are shown in Figure 3.8. Sample three is the magnesia stabilized zirconia; it performed poorly in both slips. TZP ceramics held up very well even though the hardness of SiC in the slip is much higher than that of zirconia. Amazingly, the two TZP ceramics acted differently in the alumina slip, but acted in a similar fashion in the SiC slip.

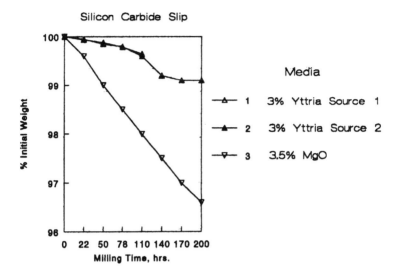

Figure 3.8: Wear-Zirconias, Silicon Carbide Slip

Figure 3.9 displays the data for both the SiC and Si_3N_4 media in the SiC slip. Again, SiC did not show detectable wear; this is surprising as the slip particles are of the same hardness. However, Si_3N_4 showed a normal wear pattern without the initial no-wear zone seen in the alumina slip results. While this is perplexing, one can at least measure the wear rates to select the grinding media. This type of test can be scaled up and is commonly used on larger media.

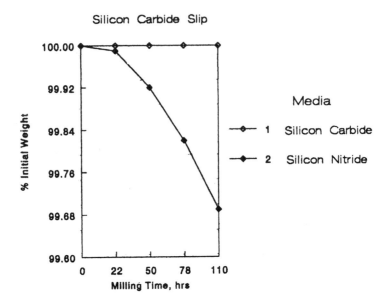

Figure 3.9: Wear-Silicon Carbide/Silicon Nitride, SiC Slip

The wear rate is a function of media size. Smaller diameter media wear faster than larger diameter media. Figure 3.10 shows data on the TZP media in the SiC slip (from the TOSOH manufacturer).

The curves are linear with milling time. There does not seem to be a quantitative relationship between the wear rate and the media diameter in this test. There are other factors other than surface area or mass that should be considered here.

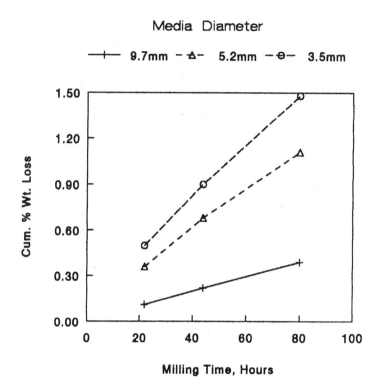

Figure 3.10: Wear of TZP vs Media Diameter

Data using commercial spherical grinding media of the modified fused zircon is shown in Figure 3.11(Courtesy of SEPR). This material has the advantage of being cheap relative to other higher purity types. It is also durable in many applications when compared to those of severe attrition. The microstructure is of interest as it suggests a reason for the good wear resistance.

Figure 3.11: SEPR Beads (Courtesy of SEPR)

The microstructure of the SEPR beads is shown in Figure 3.12. Zirconia dendrites are seen in the microstructure radiating out from central sources. The matrix is a glass with silica and other constituents added during manufacture.

The TZP media is much more wear resistant under highly innocuous conditions. This media has a pearly luster due to a fully dense microstructure and a very high index of refraction. When this happens, the wear mechanism is on a molecular scale producing the high polish. This polish is retained on the media surfaces. These TOSOH beads are shown in Figure 3.13 (Courtesy of TOSOH).

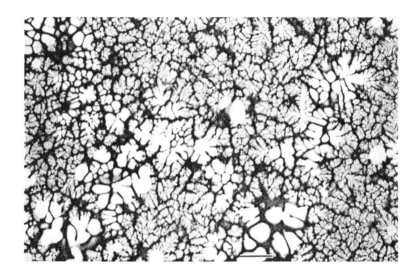

Figure 3.12: Microstructure of SEPR Beads in Polished Section

Figure 3.13: TOSOH Beads (Courtesy of TOSOH)

6.0 MILL RACKS

Laboratory jar mills are usually rotated on a mill rack. The rack is a steel frame with two rollers per tier with usually one to three tiers. On each tier, there are two rollers: a drive roller and an idler. The drive roller is usually driven with a chain and sprocket connected to a gear reduction motor. Since mills come in different diameters, it is advantageous to have a variable speed drive and a tachometer for each tier. The drive roller must have a force vector pulling down on the mill body, as shown in Figure 3.14.

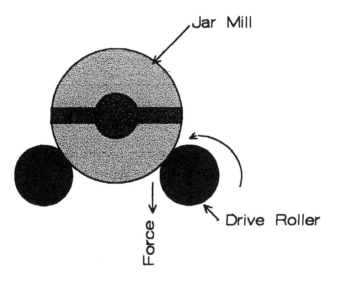

Figure 3.14: Mill Roller Direction

When the drive roller force vector pulls up, the mill will slip and not roll at the proper speed. The proper speed is 60-70% of the critical speed. As the mill rotates, the grinding media is subjected to two accelerations: gravitational force and centrifugal force. When these are equal or when centrifugal acceleration is the greater, the media will just ride around and not do any grinding. This limiting factor is the critical speed given in the equation below:

$$a = v^2/r \qquad\qquad (3.1)$$

Where a is the acceleration due to rotation, v is the velocity around the mill's internal diameter (ID), and r is the mill's ID. In most labs, there are mills of different internal diameters, therefore the mention of a variable speed drive. For each diameter, there will be an optimum rack speed. In design, there are six important factors: a gear ratio of the mechanism, a gear motor in the right speed range, a variable speed drive, a separate speed control for each tier, a tachometer for each tier, and a timer for each tier. The idler roller should be adjustable laterally to accommodate different diameter mills. Commercial mill racks usually do not have all these features, but they can be added if requested.

There are two types of rollers: steel and rubber. For the mill to rotate properly, there should be enough friction to turn the mill body. Rubber against steel is satisfactory, where either the rubber or the steel can be on the mill or on the roller. Steel does not have enough friction against either rigid polymers or itself to turn the mill. Rubber rollers are available with molded on-end flanges that are a great help to keep the mill on the rack. Jar mills will always migrate to the end of the rack, where they will fall off without the flanges or other restraining mechanisms.

With more than one mill on the rack, the jar mills will all migrate to one end and will rub against each other. As a result, either the closure mechanism on one mill can flip the other off the rack, or the closure on one mill can unscrew the closure on the other. If this occurs, the batch will spill over the mill-rack, onto the lower tiers, and onto the floor where it will harden. This in not an uncommon phenomenon as the jar mill diameters are different, causing the end closures to get entangled. Additionally, the mills are rotating at different centers and at different speeds; causing the closure screw to unwind, the lid to open and the batch to spill.

The temperature of the mill batch can increase due to the friction in the mill. This in turn will cause the internal pressure to increase and possibly cause the batch to leak out of the mill.

Since each ceramist's experimental needs vary, multiple users can further complicate the management of these problems. These problems are manageable if only one person uses the mill rack at a time.

There is an additional problem associated with the jar mills. Layers of dried slip can build up on the lid, gasket, and top of the mill body. This buildup makes it difficult to maintain a leak-proof seal. Additionally, as the bolt in the cross bar clamp is repeatedly tightened, the cross bar bends into an arc and no longer fits the ears on the mill-body. This in turn can cause the slip to spill. Keep the lid, gasket, and top of the mill body clean. Hardened slip can be hard to remove and a scraper is often needed.

To obtain the desired particle size, the milling time is specified. Any change in that time will produce a different particle size distribution with a coarser size when the time is shorter and a finer size when the time is longer. Problems arise whenever there is a power failure or when someone turns off the mill rack to remove a mill. With the mill rack off, the time is compromised. It is not so bad if this is evident, but it is a serious problem when not evident. There has to be an understanding among people using the mill rack that the speed cannot be changed or the rack left off other than momentarily when someone else has a mill running. It would be wise to measure the particle size distribution. If it the particle size is not reasonable, one should start over.

A commercial mill rack is shown in Figure 3.15 (Courtesy of E.R. Advanced Ceramics). This mill rack has a variable speed, but the speed is the same for both tiers. This is not a problem provided the mills are all of the same diameters.

Figure 3.15: Commercial Mill Rack

Mill racks are noisy and are generally placed in an enclosure with a door in front. A mill enclosure helps reduce noise that emanates from the mill rack, but it also makes it difficult to clean the spills and to move the mill due to its weight. Figure 3.16 is a schematic of such an enclosure.[4]

The enclosure is shown for a single tier. Since the lid on the enclosure is double hinged, it is lifted away providing easy access to the rack and a better angle. This enables one to lift the mill using one's legs and not one's back.

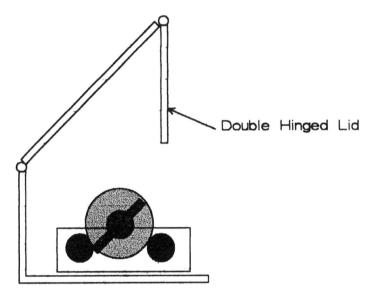

Figure 3.16: Mill Rack Enclosure

7.0 OTHER MILL DESIGNS

Some mill designs will not be discussed as they are not usually seen in the ceramic laboratory. These mills are too large for lab use, the batch can be easily contaminated, and clean up is very difficult. In a plant, the mills repeatedly run the same formulation and clean up is not a problem, but clean up is essential in the lab as many different formulations are used. The following five mill types will be discussed: attritors, stirred mills, vibrating mills, planetary mills, and jet mills.

Attritors

Attritors have a cylindrical shell that can be lined with wear-resistant materials and a rotating vertical impeller. The impeller has cross bars that can be covered with wear-resistant, ceramic sleeves. The attritor is filled about two-thirds full with the grinding media. All stirred mills should use a spherical media as other media shapes tend to over pack and overload the drive. Stirred mills are much more energetic than ball mills; as such, milling times are reduced. Seals and bearings are up out of the batch creating a maintenance advantage. Attritors can be operated in three modes: batch, recirculating, and continuous sometimes in series. Most lab work uses the batch mode as shown in Figure 3.17 (Courtesy of Union Process)

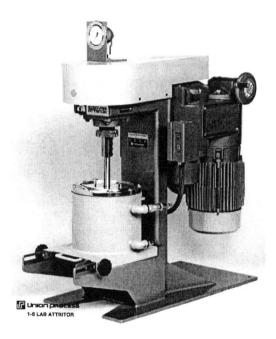

Figure 3.17: Attritor (Courtesy of Union Process.)

Stirred Mills

The stirred mills, like the attritors, have a grinding media agitated with a mechanism such as an impeller or a rotating cylinder. Some of these mills have a horizontal impeller that places the seals and bearings into the batch. Some stainless steel impellers wear and contaminate the batch. Milling equipment develops and evolves so fast that this discussion will likely be obsolete by the time it is read. Nevertheless, the principles will have a longer lasting value.

Epworth Manufacturing Company

A horizontal stainless shaft places the seals and bearings into the batch and the stainless wear. This is a high intensity mill that requires a minimal amount of media.

Welte RS Pulversizer

The rotating cylindrical part can be coated with a wear-resistant surface. Seals and bearings are up out of the mix. This mill has minimal media requirements, and wear resistant liners can be used.

Netzsch Small Media Mills

These mills provide options on the mechanical action and minimize media requirements. Milling is done in an annulus with the media. The rotor is horizontal and places the bearings in the mix; however, a mechanical seal keeps the seals flushed.

Turbomill

This Turbomill has a rotating slotted cylindrical cage that is constructed of high density polyethylene.[5] Since the cage wears during milling, it adds a considerable amount of polymeric contamination to the batch. The cage is vertical; hence, the seals and bearings are down in the mix.

Cyclomill

This Cyclomill design is experimental; it is included here as it may have novel advantages.[6] The mill gyrates rather than rotates. To visualize the motion, think of a Ferris wheel but substitute the chairs with cylindrical, grinding vessels. An increase in the gravitational force is due to the gyration. A potential advantage is that there are no seals or bearings in the mix. With the use of flexible tubing, the mill can be set up for continuous grinding. The mill can also be lined, and it uses a modest quantity of grinding media.

The stirred mills grind faster than ball mills. Additionally, they use less of the grinding media. (This is an issue as high quality media are expensive.) The 2-3 mm diameter, TZP spheres are preferred in many cases. Most stirred mills are water jacketed to disperse the heat. Without a heat dispersal mechanism, one could overheat the binder.

Vibratory Mills

Vibratory mills usually consist of annular-shaped tubs mounted on a shaker base.[7] The tub is about two-thirds full of a grinding medium that can be spherical or cylindrical. Cylinders with rounded ends and with 1-1.5 cm diameters are commonly available. An adjustable, off-balanced rotor imposes the vibration. One can use a batch mode or a circulating mode of operation, though the batch mode is commonly used in the lab. To minimize contamination, the tub can be lined with a polymer lining. The milling intensity is moderate and is somewhere between the intensity of the ball and stirred mills. Vibratory mills are good choices for pure ceramic powders due to the high purity media and the low wear rate. In one test, MgO stabilized zirconia showed a high wear rate; it added 1-2% ZrO_2 to a Bayer alumina batch during milling. This illustrates that the wear rate of the media can be very important.

Planetary Mills

Planetary mills usually consist of four, small, balanced mills in a planetary array. This setup helps achieve higher gravitational forces that help shorten the grinding time. Gearing imparts two rotary motions: around the central axis and around the axis of each smaller mill. For some reason, these mills are not found in many laboratories. It could be because of the four small mills that have to be batched, balanced, and cleaned.

Mill Clean Up

For purposes of this discussion, one assumes that the type of application selected is contingent on minimizing contamination. Of course, this is not always true. One should also consider what is more appropriately necessary in achieving the required result.

Cleaning the Milling Media

Empty the contents of the jar mill onto a screen placed over a container. Vibration will enhance recovery of the slip as most slips are pseudoplastic. One can construct a rig to hold both the screen and slip receptacle on the vibrator. When dealing with viscous slips, spray a mist of water onto the grinding media to help recover the additional slip. The screen containing the media is rinsed with water and then ultrasonically cleaned with a detergent. Rinse the media with deionized or distilled water, place the media back into the mill with pure water and then roll the media. Repeat this cleaning process before finally oven-drying the media. It will take about two cleaning cycles for the water to run clean.

The cleaning process is exacerbating when it is not practical to remove the grinding media from the mill. An option is to repeatedly run pure water in the mill until it runs clear. Circulating milling systems have additional complications with pipes, fittings, tubes, seals, pumps, and tanks. With the initial assumption that the utmost cleanliness is a prerequisite, the entire system will have to be dismantled and cleaned piece by piece.

8.0 BLUNGERS

A blunger is an impeller that vigorously stirs the slip in a tank. Ordinarily, one would think of it as a mixer rather than milling equipment, but there is an exception. Clay slips are often deagglomerated by blunging with a surfactant. Laboratory blungers can be small, stirring motors with a shaft and a propeller. A blunger could also be a mixer such as is found in the kitchen.[8] Additionally, it could be a variable-speed drive with a vertical shaft ending in a disc-shaped impeller. Such a mechanism moves up and down with the help of an air cylinder. A blunger of this type is shown in Figure 3.18.

The intense shear produced in the slip depends on the impeller's peripheral velocity and the slip viscosity. The impingement between rotating particles in the slip rather than the impingement on the impeller causes the deagglomeration. Deagglomeration is enhanced when the slurry contains coarse particles as they act as a grinding media.

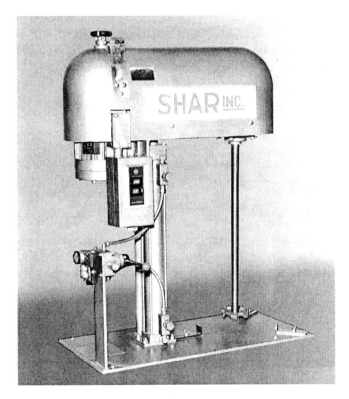

Figure 3.18: Blunger (Courtesy of Shar)

Check List

The following check list helps summarize the contents of this chapter.
- Jar mill type: porcelain, high alumina, polyurethane, neoprene, steel, and high purity ceramics
- Contamination from the mill
- Dead spots
- Dumping procedure

- Media: type, size, shape, specific gravity, wear rate, fill, and testing procedures
- Mill racks: rotation, variable speed, critical speed, tachometer, idler adjustment, roller/mill friction, containment flanges, mill accidents, enclosures, and timing problems
- Other mill designs: attritors, stirred, vibratory, jet, and planetary
- Clean up
- Blungers

REFERENCES

1. Alan G. King. "Improved Design for Milling of Advanced Ceramics." Bull. Amer. Ceram. Soc. 72[12] 43 (1993).
2. Sweco Process Industries
3. SEPR
4. Personal Communication, Don Fink
5. Dale E. Wittmer. "Alternative Processing Through Turbomilling." Ceramic Bulletin, 67[10] 1670-2 (1988).
6. Lawrence Gamblin, personal communication.
7. Sweco Process Industries
8. Hobart

4

Slip Preparation Procedures

1. 0 INTRODUCTION

Slips are a suspension of ceramic particles in a liquid that is usually water. Slips also contain other materials such as deflocculants, pH modifiers, and binders (if the dried slip is to be used for pressing). There are two principal uses for slips: slip casting and making press mixes. Slip casting is the process where the slip is poured into a plaster mold and the water is drawn into the plaster, creating a wall of consolidated material. This can continue until the part is solid, or can be interrupted to make a thin walled part. Both fine and coarse grained slips are discussed in this chapter.

While arbitrary, in this discussion, fine slips have a grain size of between 0.1 micrometers and five micrometers. Coarse grained slips have grains up to about two thousand micrometers (eight mesh), with added fines that act as the bonding agent after sintering. These dimensions bracket the discussion in a loose way. Both fine and coarse slips have much in common as to processing procedures, with some differences that will be discussed. Generally, slips are prepared by the following methods: selecting materials, batching, incorporating additives to control rheology, milling or mixing, filtering (fine slips), de-airing (usually fine slips), and storing.

2.0 SELECTION OF MATERIALS

Selection of the starting raw ceramic materials is crucial. It is virtually impossible to obtain the optimum properties if the powders or grain do not have the right characteristics. Both fine and coarse materials will be discussed.

Fine Grained Powders

Four attributes of powders will be discussed in relation to the selection of materials. These are particle size distribution, shape, purity, and flaw sources.

Particle Size Distribution

There are two conflicting consequences to particle size in this range (0.1-5.0 μm). These are sinterability and processing characteristics. Very fine particles have a high surface area that can be used to lower sintering temperatures, increase fired density, and produce a small grain size in the fired ceramic. These ceramics are wear resistant and strong. Toward the lower end of the range, the amount of surface makes these powders strongly influenced by absorbed materials and inter particulate forces. They become difficult to process and result in a low green density and a high firing shrinkage.

The surface area is much lower toward the upper end of the size range.. This requires a higher firing temperature, a lower firing density, and a coarser grain when fired. These ceramics are easier to process since surface effects are much less. Typically, the green density is higher and the firing shrinkage is less. Alumina, in the upper part of the size range, can have a glassy bonding phase.

In the lower middle of the range (0.3-0.6 μm), a compromise is reached where processing is manageable. Sintering temperatures are relatively low, firing shrinkage is fair, fired density is near theoretical, and grain size in the ceramic is small (1-3 μm).

Fracture Sources

Strength of fine ceramics is sensitive to flaws in the structure, which serve as fracture origins. Common defects are agglomerates or extraneous materials in the powder. Internal flaws in the structure are sources of fracture origins as seen at two magnifications in Figure 4.1.

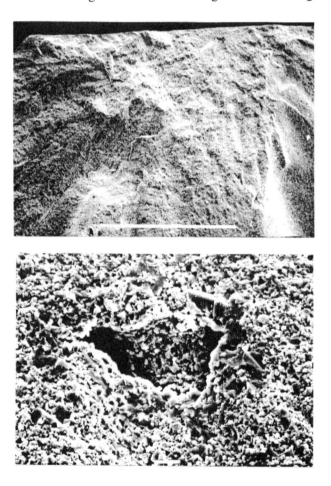

Figure 4.1: Flaw Origin on a Fracture Surface. The arrow points to the fracture origin. Scale bars: A 1000 μm, B 10 μm.

The material is a TZP ceramic with a void as the fracture origin. The arrow in the figure points to the flaw. Agglomerates are more common origins. The relatively smooth area around the origin is called the mirror; it often helps to locate the origin. At higher magnification, the origin is seen to be a 50 μm void that concentrates stresses. This results in the fracture at a lower strength than the ceramic would have shown if the void was absent.

Consider a tensile test specimen measuring 1 cm x 1 cm x 10 cm. Assuming the grain size to be one μm, there will be about 10^4 x 10^4 x 10^5 grains in the structure. This multiplies out to 10^{13} grains, where one inclusion 10 μm in diameter can compromise the measured strength.

Figure 4.2 shows the effect of flaw removal on the hypothetical strength distributions.[1]

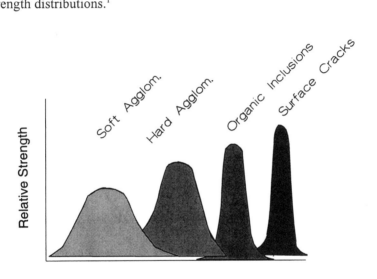

Strength Distributions

Figure 4.2: Effect of Flaw Removal on Strength. Strength increases and becomes tighter in distribution as flaw sources are removed.

The curves are conceptual but they propose a point of view that has merit. The author took the liberty of redrawing the curves in the reference, since the precision of the measurements can increase as the flaws are sequentially removed. This is seen conceptually in that, while the strengths increase, the distributions also become narrower. There are four types of flaws suggested here: soft agglomerates, hard agglomerates, organic inclusions, and surface cracks. The first three result from the starting powder. As each type of flaw is removed, the ceramic becomes stronger and has a tighter distribution. This strategy can be made to work for fine grained ceramics, since substantial progress can be made by removing flaw populations. Milling, classification, calcining, and polishing or annealing techniques are applicable in this regard.

There are very few techniques to control the particle size distribution with these fine powders. The coarse end of the distribution can be truncated by classification. Classification is a process for narrowing the particle size distribution. One such classification process is settling. The settling process, however, is hard to scale up since a large settling pond is needed for volume, and it is difficult to keep the pond free from contamination.

The fine end can also be reduced by sedimentation depending on the size of the particles. Five μm particles will settle out in a reasonable time allowing decanting removal of the fines. Air classification is another technique that will strip off the coarse end of the distribution at the higher end. Fine grained particles are much more difficult to classify since energy from the fluid does not couple efficiently with the particles. The powder supplier provides the distribution. Milling to break up agglomerates is usually the practical technique for processing.

Aspect ratio

An aspect ratio is the ratio of length to width. Two cases where the aspect ratio is crucial are fibers and platelets.

Fibers are difficult to incorporate uniformly in a batch as they form a tangle, with the bonding powder sifting through. A tangle will also prevent sintering to a high density, and the ceramic may have to be hot pressed such as SiC whiskers in an alumina matrix.[2]

As the particle size decreases, the surface area increases by the square of the radius. Surface chemistry dominates slip rheology; thus making surfactants, solids content, pH, and electrolytes the prominent players. Anything affecting the surface will have a dominant effect. The details of the adsorption layering control rheology. Because of this, it makes a difference about which surface active agent one applies first. This will be discussed later.

Chemical Purity

Fine grained ceramics usually contain powders ranging from 98% to 99.999% purity. For optimum properties, one uses a purer powder. The impurities have three choices: to diffuse into the grains, form separate phases, or to attach to the grain boundaries. Those that diffuse into the grains are not particularly harmful to mechanical properties but may affect optical or electronic properties. Separate impurity phases are often fracture origins that lower the strength of the ceramic. Grain boundary phases are often glassy and will lower the creep resistance of the ceramic. For every generalization, there is always an exception. In this regard, there is the alumina/yttria/silica glass in Si_3N_4 that is necessary for the ceramic to sinter. Alumina ceramics often use a Bayer alumina and some of them contain as much as 0.4% Na_2O (soda). This extra soda forms a glassy phase with other impurities on the grain boundaries.

Check List, Fine Powders

When selecting and working with fine powders, refer to the following list:
- particle size distribution
  - Less than 0.3 μm - difficult to process
  - 0.3-0.6 μm - preferable for fine ceramics
  - 3.0-5.0 μm - preferable for conventional ceramics,
- flaw removal for fine ceramics,
- equiaxed particles for most applications,
- high purity powders for fine ceramics,
- lower cost powders when necessary, and
- selection of the starting powder crucial in many applications.

Coarse-Grained Materials

Coarse particles are often called grog. While the term is not exact, it generally refers to eight mesh and coarser. One can obtain coarse grogs in two ways: sintered or fused that are then crushed and sized. Sintered grain has internal grain boundaries making it less stable dimensionally at high temperatures. Since the grain sintering temperature is higher than the use temperature, the sintered grain is fairly stable. Fused grain often consists of single crystal domains, which makes it thermally stable. Fused grain is often more jagged than sintered, making it more difficult to pack densely. Particle packing is an issue with coarse materials, because the ceramics and refractories have little firing shrinkage. The green density is near the fired density. Packing is a result of particle size distribution and the aspect ratio.

Particle size Distribution

There are three lines of thought on packing strategies. For fine materials, one thought is to have spherical particles all of the same size.[3] Also, for porous ceramics, a limited range of coarse sizes controls the v/o% of pores, pore size, and permeability. The second strategy is to gap size the mixture where the sizes fall into a sequence where each smaller size just fits into the interstices of the next coarser size; the ratio of diameters is about 10:1. The third strategy is to blend two or three natural distributions. These are designated by the top size and finer (F). One such example is 24F where all the material passes through a 24-mesh screen.

Except for porous ceramics, this latter strategy of distribution is most economical and practical. This will be discussed in the section on Coarse Particle Sized Slips.

Flaws

The coarse grains themselves act as crack sources and since one does not introduce other damaging things in the processing, flaw populations are not a concern. It is unwise to assume that there will not be strange things in the grain, so it is a good idea to examine the grain at least with a stereo binocular or a petrographical microscope looking for extraneous materials.

Aspect Ratio

Coarse-grained particles are the result of crushing, leaving jagged particle shapes. As these do not pack very well, it may be necessary to mull the grain. Iron is introduced and may have to be removed. Dry magnetic separation is inexpensive. When this is not good enough, spend more and use an acid leach.

Fibrous materials and platey shapes are also used in refractories and alumina ceramics. One example is talc in a cordierite batch. Talc is a common source of MgO in ceramics and refractories. Platey shapes pack well, like bricks in a wall. Fibrous shapes do not pack well; they form tangles.

Platelets are more commonly found in traditional ceramics and refractories than they are in finer ceramics. Graphite, coarser Bayer alumina, micaceous minerals, and clays are platey. Plates will orient themselves during processing and will also segregate, especially when their size is large. Sometimes orientation is needed, an example being foundry crucibles where graphite platelets can be oriented parallel to the crucible wall, thereby decreasing metal penetration and thermal conductivity.

Whitewares are mostly clays reconstituted into the glassy phase during the firing process. Clays and related platey materials are slippery and sometimes are added to help in the processing.

Check List, Coarse grain

When working with coarse grains, observe the following list:
- choose a sintered or fused grain,
- select gap sizing or blending of normal distributions, and
- choose a grain shape, mull if a higher density is required.

3.0 FINE PARTICLE SIZED SLIPS

This section addresses two aspects of these fine slips: particle size measurements and the physical constitution and chemistry effect on viscosity.

Particle Size Measurements

One technique for measuring particle size is microscopy. By our definition, the fine particles are too small for optical microscopy to be of much help for measuring size. So, we need to use electron microscopy. There are two main types: scanning and transmission microscopy.

Scanning Electron Microscopy (SEM)

The useful range of magnification is enormous, ranging from about 4X to about 20,000X. Resolution limitations, especially for electronic insulators, start to blur the image at extreme magnification. Depth of field is large because of the scanning method that samples one layer at a time and then assembles them into the field of view. There is no color except artificial, and there is no access to optical properties. Wave length dispersive or energy dispersive analytical techniques determine the chemical composition of the particles. The capacity to obtain compositional information on the surface is a valuable asset for the SEM instrument.

Preparation of the sample may be difficult. One strives to obtain a uniform dispersion of the particles on the sample holder, called a stub. The particles may clump up as the droplet dries due to surface tension. Dispersions in some organic liquids can produce an improvement. Acetone or low molecular weight alcohols have lower surface tension and a lesser tendency for capillary forces to agglomerate particles on the stub. After the sample is coated by sputtering with a metal to make it conductive, it is placed in the SEM. The analysis centers on particle shape, size, agglomeration, and phase composition.

There is no way that one sample can be truly representative of the material because it is so small. Examination of additional samples increases the confidence in the measurements, but this is time consuming and expensive. A good practice is to examine at least three fields of view.

Whenever there is more than one particle composition, energy dispersive analysis will identify each different particle composition. It is useful and advisable to have an energy dispersive capability in the microscope. Wave length dispersive capability can analyze particle compositions to atomic number 5 (Boron) and a sensitivity to parts per million for some elements. Wave length dispersive capability is more difficult to use, but it may be needed to solve certain problems. When an SEM is acquired, both energy and wave length dispersive capabilities should be available as options that can be added later.

The SEM photos can be analyzed for making measurements of particle dimensions. The microscope shows the magnification on the screen. If there is the need for exact measurements, latex spheres in a few sizes are available for such a calibration. One measures only two dimensions in the plane of view. Image analyzers are available when there is an extreme amount of work to do. For most analyses, a good ruler is sufficient to make the measurements. The technique is to place randomly spaced lines on the photo and, without regard to judgment, measure the intercept of each particle. Particle size distribution will converge quickly. Figure 4.3 is a schematic view of this measurement process.

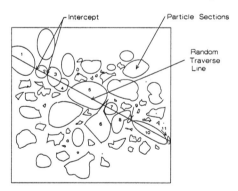

Figure 4.3: Particle Size by Intercept Measurement. Randomly spaced lines are used to measure each particle.

The field of view is transected by a randomly located line. In this illustration, there are 11 numbered particles along the line. Each particle's dimension is measured along the line intercept as shown in Table 4.1.

Table 4.1: Particle and Intercept Measurements

Particle Number	Intercept (mm)	Comment
1		Off the field of view
2	10	
3	10.5	
4		Edge covered up
5	27	
6	2	While deceptive, it will even out
7	8	
8	8	Do not be concerned about elongation
9	2	Measure small particles
10	29	Elongation does not matter
11	1.5	First measurement completed, move line another random location

While the above data is hardly enough, the mean size and standard deviation can be calculated. The mean is 10.8 mm (divide by the magnification), and the standard deviation is 10.3 mm. The mean and standard deviation will converge quickly. An additional five lines were drawn across Figure 4.3. The resulting data, taken with a ruler, is illustrated in Figure 4.4.

The curves are cumulative as data is added. The mean levels out very quickly, while the standard deviation continues to drop at a low rate.

There is a total of only 29 measurements represented in these curves, with reasonable results. However, data from 10 lines would be more appropriate, as this only takes a few minutes.

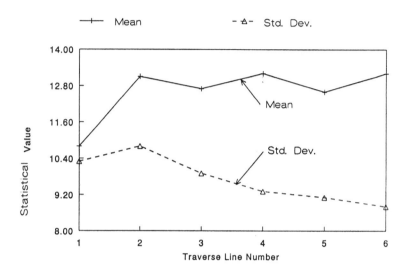

Figure 4.4: Convergence of Particle Size Data. The mean converges quickly. Standard deviation starts to level out.

SEM microscopy has an advantage in that the particle shape is seen and if other phases are present, they can be observed and measured. This technique is also useful for measuring the volume % of other phases or porosity using polished sections. One can see the particles in Figure 4.5 that is an SEM view of alumina, A16-SG (Alcoa).

One can use such a field to make measurements of the PSD. However, it would be better if the particles were more dispersed and at a higher magnification.

Figure 4.5: SEM View of A16-SG. Ideally, the particles should be dispersed evenly. Scale bar 1 μm

Transmission Electron Microscopy (TEM)

The resolution from TEM is much better than the resolution obtained from SEM. Sometimes with using TEM, one can resolve the layers of atoms in a lattice plane. TEM is much more laborious and has fallen into disuse for particle measurement applications unless the particle size is below the SEM range. Wherever it is appropriate, TEM should be considered. Figure 4.6 illustrates the detail of ceramic particles when using TEM.

The micrograph shows plasma dissociated zirconia using a replica technique. Detail is superior to other methods, but it is tedious and takes skill. Whenever this amount of detail is needed for either technical or aesthetic reasons, TEM replicas should be considered.

Figure 4.6: TEM View of Plasma Dissociated Zirconia. Resolution is excellent with TEM. Scale bar 1 μm.

There is another technique for TEM or optical microscopy that produces data on all three particle dimensions. Particles in a dispersed state can be shadowed at an angle (15 degrees) on a carbon film. This produces a photo that shows both the dimensions in the plane and, by trigonometry, the vertical dimension. Keep in mind that particles tend to lay flat when sedimented.

A shadowed TEM photo is shown in Figure 4.7. The material is silicon carbide powder with the shadows obtained by oblique (15 degree) vapor deposition. Figure 4.8 shows the geometry.

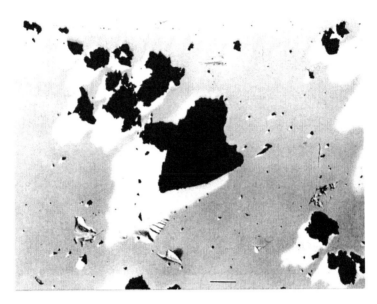

Figure 4.7: TEM Shadowed SiC Particles. Shadowing shows the vertical dimension. (Scale bar 2 μm)

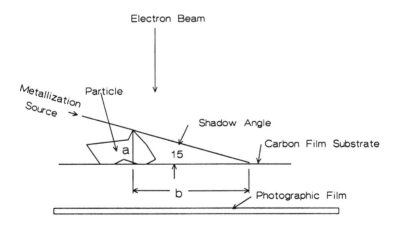

Figure 4.8: Shadowing Geometry. Fifteen degrees is a useful angle for shadowing. The particle height is Tan15° = h/1.

Data can be represented as a triaxial plot as shown in Figure 4.9. In the photo, the longer dimension (a) is in the plane of view along with the shorter dimension (b) and the calculated vertical dimension (c). The plot shows the ratio of 'a/b' along the horizontal base and the calculated height along the vertical axis. Part of the diagram is excluded either by definition or by settling mechanics. Since the particles lay flat, 'a' is greater than 'b' by definition, and 'c' will almost always be smaller than either 'a' or 'b'.

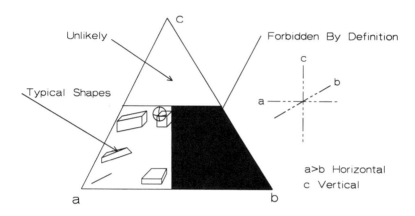

Figure 4.9: Particle Geometry Plot. The right half of the diagram is excluded by definition. The top is improbable because particles lay flat.

Some shapes are located on the diagram for reference. A plot of all three dimensions adds a third dimension to the information on particle shape.

The data as presented is quantitative. Platelets or fibers are oriented flat because of their geometry, thus establishing a built-in bias on particle shape measurement. This is one technique to obtain the vertical dimension. With larger particles, one can use the fine focus of the optical microscope to obtain the third dimension. To obtain this, one focuses on the top of the particle and then on the slide surface, with the difference being a measure of the height. Precision depends on the magnification of the objective lens. A cover glass over the particles will protect the objective lens from being scratched.

Microscopic methods for measuring particle size, especially for fine particles, have largely given way to other techniques such as light scattering (diffraction). Over the past few years, these instruments have greatly improved in quality and many good choices are available presently. I have a conceptual problem with light diffraction or scattering as a measurement tool. The diffraction spectrum is diffuse with only a few broad bands. To put it another way, no crisp data is available in the spectrum. It is a stretch to calculate the particle size distribution in detail from such diffuse data. Nevertheless, the data reduction is fast, convenient, and reproducible. Since the distribution is periodically checked against microscopy, the author would use this method as a matter of choice.

Alternatively, there are sedimentation techniques: centrifuging and light absorption or by settling and x-ray absorption. These are good methods for particles that are larger than about 2-4 μm. Another method is to measure the electrical conductivity in an orifice as the particles stream through. Conductivity will decrease in proportion to the cross section of the particle. There are three problems with this technique. The orifice can become plugged, the measurements are not sensitive for small sizes, and the particles can be some shape other than block-like. All could result in false readings.

Sample preparation is important. One good method involves ultra-sonic dispersion in a soundproof chamber and using a magnetic stirrer placed on a lab jack. A flat tipped ultrasonic horn works well. The end of the horn will erode in the center but not on the edges. It can be machined smooth but do not take off too much material or the horn will be off-tune. Use titanium for the horn. (The machinist should be informed of this as titanium is difficult to machine.) A stirring rotor is Teflon coated and works best if it has a slanted top surface that lifts the liquid so that all of the material is uniformly dispersed. Dispersants are commonly used. Dispersion time and intensity are standardized. Typical time could be about 3-5 minutes, and intensity about 60% of the instrument's capability. Establish both these factors by extending the time and intensity to where the particle size measurement stabilizes. Throwaway polyethylene pipettes are very useful to take out allocates.

Viscosity

Viscosity is a measure of the resistance of a slip to flow, or a liquid to shear. When pouring a slip out of a beaker, if the viscosity is high, it will pour slowly like molasses. When the viscosity is low, it will pour out like water. In both these cases, the flow rate is low and shear-like. When one increases the shear rate, by squirting the slip out of a hypodermic needle, slips will act differently when compared to being poured out of a beaker. This introduces a complication that has to be understood and coped with since it can make a large difference in subsequent processes. If the slip or paste is difficult to squirt with increasing velocity, it would also be more difficult to pump or extrude. Alternatively, if a slip or paste becomes easier to force out with increasing velocity, the material would be easy to pump or extrude. A third case is where the slip looks like a paste but becomes fluid when stirred. This is akin to latex paint. We will now discuss these three most common types of rheology.

Plotting shear rate against viscosity determines the rheology of the slip. The principal choices are pseudoplastic, dilatant, and thixotropic, as alluded to above. Just to be a little more thorough, there are also Newtonian and Bingham rheologies. Newtonian behavior is more typical of nonpolar liquids such as hydrocarbons or water where viscosity stays the same as the shear rate changes. Slips are not Newtonian. Aqueous ceramic slips may show Bingham behavior where there is a yield point that has to be exceeded before the slip flows; this may probably not be evident except at low shear rates using a very sensitive rheometer. However, this is important with low shear rate phenomena when there are problems with sag or leveling. It usually takes some time for thixotropic thickening to build up in slips. Sometimes, this period is measured in hours and could be a problem for a thick cast. Clay slips are usually thixotropic.

Figure 4.10 schematically demonstrates these flow curves. Newtonian is shear independent. Pseudoplastic is shear thinning. Dilatant is shear thickening. Thixotropic thickens with time.

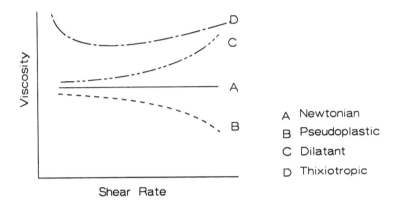

Figure 4.10: Slip Rheology Curves

Most slips are pseudoplastic in that viscosity decreases as the shear rate increases; this can be called shear thinning. The more one stirs these slips the thinner they become. Coarse grained slips are dilatant when the volume percent solid is high. A dilatant slip feels solid when poked. Tip the container and the slip will readily flow. One can observe this quality with corn starch in water as it forms a dilatant slip. Just where dilatancy occurs depends on the particle packing, which in turn depends on the size distribution and shape. Slips around 65-70 volume % solids or higher usually are dilatant.

Viscous behavior is important in ceramic processing. With fine grained slips, the following factors control viscosity: volume % solids, deflocculant concentration, dissolved polymers, pH, particle size distribution, and ionic strength. Ionic strength is generally expressed as the concentration of an electrolyte in the slip. These are the variables one can work with; a combination of them determines the slip's processing properties. Dissolved polymers are often binders that will increase green strength and aid in pressing a part. Temperature is also a factor as slips become thinner with an increase in temperature. Milling often significantly increases the temperature of the slip; hence, before measuring viscosity, the slip should first cool to room temperature.

Viscometers

Viscometers measure the viscosity of a fluid or slip as is illustrated in Chapter XI that deals with property measurements. The two principal viscometers are of the rotor cup geometry and the plate and cone geometry. Both these viscometers are suitable for fine-grained slips. When the particle size is coarse, one can only use the rotor type. Rotor size is chosen to put the measurements on scale with the instrument and with the slip. Higher viscosities are on scale with smaller rotors. Viscosity measurements are taken incrementally at increasing rotor speeds, first up and then down. A viscometer measures viscosity with rotor rpm. A rheometer measures shear stress with shear rate. From this data, one can determine the viscosity. Standards are available for viscosity and used to calibrate the instrument.

There are two types of rheometers: a shear-rate-controlled rheometer and a stress-controlled rheometer. Most rheometers are shear rate controlled, which means that the rotor speed is adjusted and the shear stress is measured. The stress-controlled rheometers do just the opposite, setting the stress and then measuring the shear rate necessary to produce that stress level. If the slip's rheology is sensitive at low strain rates, a stress-controlled rheometer is preferred. Problems with inks, silk screening, leveling, or sagging are typical applications where low strain rates are important. For most ceramic slips, a strain-rate-controlled viscometer is suitable.

The Zahn cup is a cylindrical stainless steel cup with a rounded bottom and a hole through which the slip can drain. One measures the time it takes to drain the cup. It is inexpensive and is better than nothing as a QC tool. However, it does not give information on the slip's sensitivity to the shear rate, which is an important factor.

Viscosity-Milling Time

Ordinarily, when starting with a fine powder, the powder is agglomerated and has to be milled to obtain a dispersed slip. Viscosity increases with milling due to the generation of additional active surfaces. Material within an agglomerate is not available to interact with the slip rheology. After it is released by milling, it becomes an active participant, causing the viscosity to increase. A typical viscosity curve is shown in Figure 4.11.

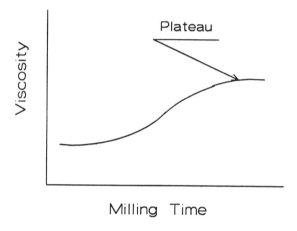

Figure 4.11: Viscosity vs Milling Time. Viscosity increases with milling time as agglomerates provide additional active surface.

The curve appears to level out with longer milling times as it becomes much more difficult to fracture smaller fragments in the mill. One questions when it is appropriate to stop milling. The answer is linked to sintering behavior, which in turn is governed by the particle size distribution (PSD), green density, sintering temperature, and soak time. This is a very good scenario for a statistically designed experiment. The independent variables are known. One can determine the dependant variables from the following properties: density, strength, toughness, or perhaps thermal shock resistance.

The particle size distribution becomes finer as the slip is milled for longer times, as shown in Figure 4.12.

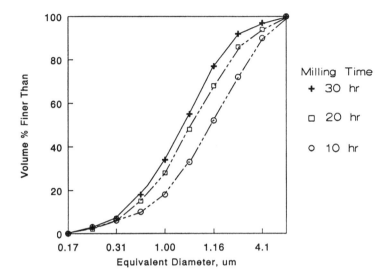

Figure 4.12: Particle Size Distribution vs Milling time. Milling decreases the particle size but more gradually as the size becomes finer.

As suggested in the Figure 4.12, it becomes more difficult to move the distribution to the left as the particles become finer. The curves tend to cluster up. When the distribution is normalized by dividing each value by the d_{50} value, all of the curves fall on top of each other showing that while the particle size is reduced the relative particle size distribution is fixed, as seen in Figure 4.13.

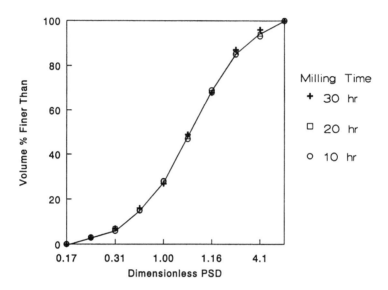

Figure 4.13: Normalized Particle Size Distribution. When each data point is divided by the mean, all of the distribution curves fall on top of each other.

The above relationship suggests that the particles in the slip are being fragmented similarly for their size as the milling progresses. There are limitations to controlling the particle size distribution. The ceramic powder can be milled finer, but the character of the distribution is fixed, at least for ball milling.

Too high a viscosity slows the milling rate. Usually, 500-1000 cps is about right. Mills are run at 60-70% of the critical speed. Critical speed is when the centrifugal force just equals the force by gravity. At that speed, the media will just ride around the inner surface of the mill and not grind anything. Ideally, the media should cascade in the mill where impacts between the media grind the particles in the batch. It is questionable when one should stop the milling process. The answer is linked to the sintering behavior, which in turn is governed by the particle size distribution, green density, sintering temperature, and soak time. These are the usual independent variables for this type of example. This is a very good scenario for a statistically designed experiment. The independent variables are as shown above. The dependant variables are determined by the following properties: density, strength, toughness, or perhaps thermal shock resistance.

Percent-Solids Viscosity

Fine-grained slips show a sharp increase in viscosity above 50 volume % solids. This is shown in Figure 4.14.

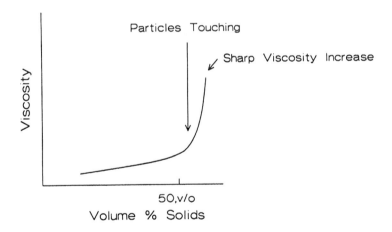

Figure 4.14: Viscosity vs. Volume % Solids. Viscosity increases sharply as more solids are added due to solid-solid contacts.

Perhaps not invariably but frequently, the volume % solids are chosen to be as high as possible consistent with viscosity restrictions. For a fine slip, 55% is high. If the slip is going to be spray dried, it is a good idea to back off a little to avoid clogging problems on the nozzle and to obtain better shaped spheres. High % solids are otherwise preferred for the following reasons: to retard segregation, to increase the casting rate, to produce a more uniform cast, and to possibly obtain a higher green density.

High solid slips need to be handled appropriately as an irreversible process called gelation could result. If the mill overheats, the organic additives will get overheated and cause the slip to gel. Addition of concentrated chemicals to the slip can also cause gelation at the location of contact where the concentration is very high. Chemicals like acids, bases, or even surfactants can cause the slip to gel at that location. It may not then re-disperse. A preferred practice is to place the additives into the water and to stir the contents before adding in the ceramic powder.

Surfactant Concentration

For slip casting, a little flocculation is often desirable to keep the cast permeable and to provide a little drying shrinkage so that the piece releases from the mold. Therefore, it is desirable to move off of the minimum of the curve but just a little to one side. The mark on the figure suggests a location. This slip will be just a little bit flocculated and will retain permeability throughout the casting. The cast will have a little drying shrinkage so it will release from the mold.

Figure 4.15 shows the viscosity vs. surfactant concentration. Viscosity drops rapidly as the surfactant is added going through a minimum and then increases again. The best concentration is often just off from the minimum where the slip is slightly flocculated as suggested by the mark.

Besides viscosity, the saturation amount of surfactant can be determined by the residual surfactant concentration in the liquid. A plot of the amount of surfactant added vs. the amount retained in the liquid shows a plateau as indicated in Figure 4.16.

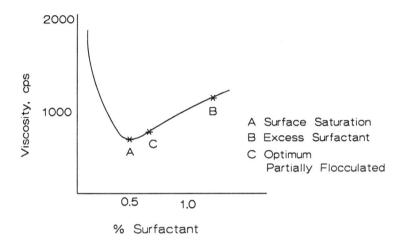

Figure 4.15: Viscosity vs. Surfactant Concentration

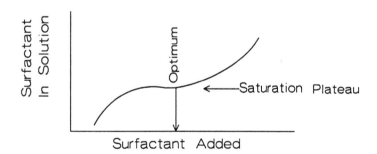

Figure 4.16: Surfactant Saturation. The amount of surfactant left in the solution plateaus as the surface becomes saturated. Plateau compositions are preferred.

When the surfaces of the particles get saturated, the residual surfactant levels out in the liquid. This is called a titration technique. One can measure the concentration using a differential refractometer or by applying other analytical techniques. Of course, standards can be made up of solutions of known properties for calibration purposes.

There are large number of surfactants that are available. Traditionally, such materials as sodium silicate, lignosulfonates, and tetra sodium pyrophosphates are available. Currently, a variety of acrylate compounds are widely used.

There is another technique to evaluate surfactant sedimentation. The method is simple but time consuming. A dilute slurry of the ceramic powder is suspended in a liquid containing a surfactant. The slurry is poured into a test tube capped to avoid evaporation, and not agitated. A well-deflocculated slip settles slowly, taking a few weeks. But when it settles it forms a hard pack. Slips that are partially flocculated soon develop a clear liquid above the slurry, but do not completely settle out in the same period. The height of the clear liquid layer and time will indicate the way the surfactant behaves.

Figure 4.17 shows the data on a sub micrometer diameter SiC powder with a variety of surfactants. Three surfactants showed diffraction banding since the particle segregation, based on particle diameter, was astonishingly perfect. Turn back the pages and take another look at the frontispiece. The best place to choose a surfactant is toward the left of the diagram, but not all the way over.

A similar experiment was repeated using isopropyl alcohol as the medium. This is shown in Figure 4.18. Surfactants are not as active in alcohol as in water. None of the surfactants showed diffraction bands. Since isopropyl alcohol is not as polar as water, the results are not as dramatic, and none of the samples showed diffraction. Since there are so many surfactants, there may be a problem in selecting a suitable one for your material. It is time consuming to perform many experiments for this selection. Personal contacts and the literature help to narrow the selection to a reasonable number.

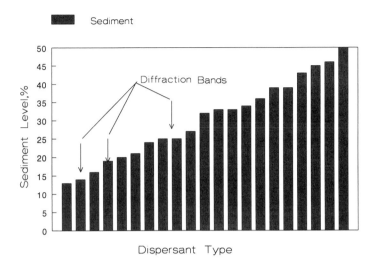

Figure 4.17: Sedimentation of SiC vs. Surfactants in Water.

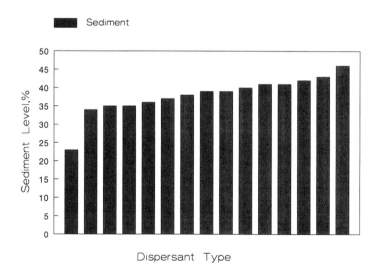

Figure 4.18: Sedimentation of SiC vs. Surfactants in Isopropyl Alcohol.

Binders

There are many binders to choose from, including waxes, alginates, starch, gums, and synthetic materials. Only a few synthetic binders will be discussed. All these binders are used for slips and include the following: polyethyleneglycol (PEG), polyvinyl alcohol (PVA), methylcellulose and its varieties, and acrylic emulsions. These four binders are manufactured by chemical processes where the composition and properties are controlled. While there can be some refinement in naturally occurring binders, they are variable in both composition and properties. Samples of binders can be obtained from suppliers in quantities suitable for lab use. Another source is from a ceramic plant.

The function of the binder in slip casting is different from its function in pressing. The density of a pressing (small pressed part) increases with the use of a binder plasticizer. In slip casting, the binder is added to increase green strength for purposes of handling or green machining. Green density is achieved by the casting process itself.

PEG. PEG is available in molecular weights from 400 to 12,000. Low molecular weights are liquids at room temperature. Useful molecular weights are from 6,000 to 12,000 with most applications at the higher end. These are waxy solids that are soluble in water, methylene chloride, and acetone. Some volatile organics require special handling, ventilation, and fire/explosion protection. Methylene chloride is toxic. Chapter Two discusses safety problems in greater detail.

PEG is a clean chemical that burns without a residue. While it is not a high-strength binder, it is adequate for most purposes. All of the binders discussed here are added as solutions or as a fluid emulsion. The reason for this is to obtain a homogeneous fluid without clumps of binder that have not completely dissolved.

The PEG solution is prepared on a hot plate using a double boiler and a stirrer. Flakes of PEG are sifted into the water bath while stirring. Ordinarily, the solution will contain between 3-5% of PEG by weight. It is a good idea to plug the stirrer into a variable voltage transformer. Stir fast enough to mix, but avoid whipping in bubbles, especially fine bubbles as they are difficult to remove. It is not a good idea to center the stir rod as this helps to form a vortex that will draw in air. It also helps to tip the stir rod at a slight angle from the vertical position as this creates turbulence. When the stir rod is centered and vertical, the binder solution will not mix well. Double boilers are available at most hardware or cook ware stores. By using a double boiler, one avoids overheating at the bottom of the pot. After the solution is completed, the pot is covered, aged overnight, and is stirred again the following day. To remove solid impurities or clumps of thick binder, the binder solution can be pressure filtered as is described in a later section. Depending on the prevailing viscosity, the bubbles will arise after a day or so. Binder solutions can be vacuum de-aired as long as the viscosity is not too high. It is a better practice to stir the binder solution into the slip and then vacuum de-air. This practice will be described in a later section.

PVA. A variety of grades of PVA make solutions over a range of viscosities. For ceramic use, one generally uses a fully-hydrolyzed binder of medium viscosity. PVA is a strong binder and is suitable and widely used for handling and green machining ceramic parts. The process for making up PVA solutions is identical to that for PEG, except that it has a higher viscosity at the same concentrations and requires more care to assure that it is homogeneous. It is a clean chemical that burns without a residue.

Methylcellulose and varieties. These products are available in a variety of grades. Variability is achieved by molecular weight, molecular weight distribution, and degree of substitution of methoxyl and hydroxypropoxyl groups. One can simplify the process of selecting a binder by phoning the supplier for additional information. This binder is unusual in that its viscosity increases as the temperature increases. Unlike the solution procedure for PEG and PVA, methylcellulose is dissolved cold. This can be a useful property that has found application in extruding honeycomb structures through a heated die, resulting in a stiffening of the mix that then retains the honeycomb structure. One can obtain a wall thickness as thin as 0.010 inches.

Methylcellulose is not as clean a chemical as PEG or PVA There is a burnout residue that is a problem only when the ceramic is sensitive to foreign impurities. Unlike PEG and PVA, it is not a strong binder and is not really suitable if the part has to be green machined. Otherwise, it is strong enough to impart handling capabilities.

Acrylic Emulsions. These emulsions are strong binders and burn out cleanly. They are useful for press mixes as well for slip casting. Being emulsions, they can break or gel. This does not infer that they are difficult to use if handled properly. As a container will be used for a long time, a crust will form on the jar lid and top. Pieces of this crust can get into the slip creating voids in the part when sintered. The emulsion can also gel in its container forming clusters that float in the emulsion. These problems can be avoided by taking allocates of the emulsion of a convenient volume and sealing them in individual glass vials. A polyethylene liner in the lid is inert and will make a good seal. When needed, one can extract a measured amount of emulsion and discard the remaining emulsion and vial. This may seem wasteful but if the allocate is not too large this procedure is much less expensive than a ruined part. Other liquids in the ceramics lab can have this same problem, and one can follow a similar procedure.

Rohm and Haas introduced an acrylic binder, Duramax[tm] B-1031 that has a high green strength. Due to the high green strength, this binder is recommended for green machining applications. Duramax is compared to PVA and PEG in Table 4.2.

Table 4.2: Binders and Green Strength

Binder	Green Strength (MPa)
Duramax	6.5
PVA	0.7
PEG	0.4

pH/Zeta Potential

The viscosity of the slip is related to the charges on the surface of the particles. Charged surfaces, when of the same sign, repel each other and disperse the slip, reducing its viscosity. Measurements of the zeta potential with pH result in a plot such as shown in Figure 4.19.

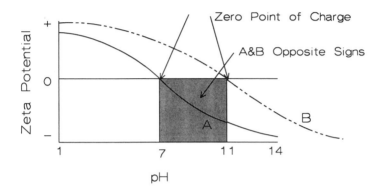

Figure 4.19: Zeta Potential vs. pH. Two materials are shown with different ZPC. A has a zero point of charge at pH 7. B has a ZPC at pH of 11.

An S-shaped curve is typical. The point where the curve crosses the zero (0) axis is the zero point of charge (ZPC). This is where the slip flocculates. For aluminum oxide, this occurs at a pH of about seven. Add either an acid or a base and the slip will liquefy. Figure 4.19 shows curves for two different materials. Material A has a ZPC at a pH of seven. Material B has a ZPC at a pH of 11. Below a pH of seven both materials have a positive charge and will be dispersed. Above a pH of 11, both materials have a negative charge and will be dispersed. Between a pH of 7 and 11, the different materials will have opposite charges on their surfaces and will attract one to the other. This phenomenon will increase viscosity, however it will also inhibit segregation of the phases. Another complication to consider is that at a pH of 7 material A will floc, while B will still be dispersed. The parallel but opposite relation occurs at a pH of 11. These details in the surface chemistry enable the ceramist to control the structure of the slip and the final sintered ceramic.

Since there may not be a continuing demand for zeta potential measurements, purchasing the instrument may not be justified. Work can be contracted out where one will pay for such measurements.

For normal laboratory work, a pH/viscosity plot can be easily created. A typical plot is shown in Figure 4.20.

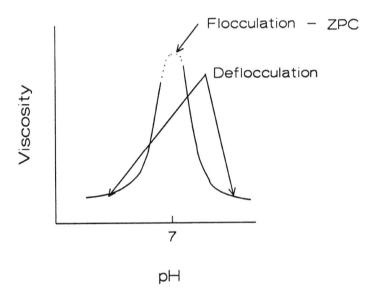

Figure 4.20: Viscosity vs. pH. Viscosity increases sharply at a specific pH for each material. The slip is deflocculated on each side of the maximum as suggested by the arrows.

A peak in the curve is typical for alumina and relates back to the zeta potential curve. The viscosity curve is more easily available however.

So far, there are four factors that affect the viscosity of the slip:

particle size, % solids, surfactant, and pH. It is a good idea to adjust the pH and to add the surfactant to the water before adding the powder. The pH will change when the powder is added. A common practice is to add the binder toward the end of the milling or after the milling. The reason for this is that the binder can be mechanically or thermally altered during milling. Adding, the binder at this point can alter the rheology. A series of trials is often necessary to derive the optimum composition. When considering a statistically designed experiment, this is a good place to zero in on the optimum formulation.

Settling

With fine slips, settling is not too big a problem. It becomes a problem when one stores the slip over an extended period. If the viscosity is low and the casting time long, there could be settling of the particles. In this case, one may need to adjust the deflocculation and viscosity.

Check List, Fine Particle Slips

In dealing with fine particle slips, one needs to pay attention to particle size measurements and viscosity factors as listed below.

- Particle size measurement techniques
  - Sampling and dispersion
  - SEM, TEM
  - Light scattering
  - Linear intercept measurements
  - Vertical measurements
- Viscosity
  - Rheology
  - Choice of viscometer type
  - Effect of milling time
  - Particle size distribution/milling time

Mill critical speed
Effect of solids v/o on viscosity
Type and amount of surfactant, saturation
Settling
Choice of binder, green machine
Preparation of the binder solution
Measurement of pH, Zeta potential

4.0 COARSE PARTICLE SLIPS

There are differences in the processing of fine and coarse particle slips. To chart these differences, the same topics will be discussed as in the previous section.

Particle Size Measurements

Screening is one technique for measuring coarse particle size distributions. We will also discuss the topics of contamination and sieving.

Screen Selection

Lab screens are usually 8 inches in diameter and a full height. These screens are available in sizes from a 4-mesh screen to a 600-mesh screen. Anything below a 325-mesh screen is difficult to use because of blinding where particles become embedded in the sieve openings. Stainless steel mesh with a brass body and brass pans and lids are preferred. Stainless steel mesh is much more durable than brass because of its greater strength. When working below a 40-mesh size, a stainless steel screen is a preferred screen to use. Above this 40-mesh size, one can use coarser mesh size screens. Fine mesh brass screens are easily damaged. A 200-mesh screen with a hole is no longer a 200-mesh screen. One can repair a screen damage by soldering a split back onto the brass body. A hole can be plugged with a drop of epoxy or solder.

One can damage the screens while cleaning them after use or by accidentally pushing a mix through the screen with a scraper. The body of the screen should be brass as they do not gall and are easier to handle. Stainless screen bodies will gall when stacked. Grit, that sifts into the space between screens, makes it difficult to separate the screens. With time, prying tools will be needed to separate the stack. Often, the stainless lip on the screen will become deformed.

Contamination

It is difficult to remove particles that become wedged between the wires in the mesh. Since stainless is harder and more durable than brass, it is a little easier to clean. With screens to about a 20-mesh size, an awl can be used to singly pop out embedded particles from the back side. Fine particles can be partially removed by ultrasonic cleaning with a detergent. One can also brush the back of the screens with a fine brass brush. This could ruin a fine mesh brass screen. When contamination is not permissible, a separate set of screens should be reserved for each material. This is not as bad as it sounds considering that one will be processing only a few materials and will only need a few sizes. One should mark these screens to identify the material before locking them away. Keep a general purpose set for borrowing or general use. A storage rack with each slot marked as to size is a good way to keep the set in order rather than being scattered throughout the facility.

Sieving

A sieve stack consists of coarse sieves on top with progressively finer sieve sizes below. One should not forget to include the lid and pan. One then places the stack into the shaker and tapes the edges. When the sieves get old and have stainless bodies, they will rust. Sieve shakers can shake, tap, or vibrate. Vibrating shakers have controls for time and intensity. These controls are set for standard conditions that can be determined by a test. When the size distribution starts to level out, that time and intensity can be your standard. Sieve shakers are noisy so a sound enclosure is a good idea. Some larger shakers have rubber balls that bounce against the bottom of the screen and reduce blinding, which can be a troublesome problem.

Viscosity

Viscosity is not as closely controlled for coarse slips as it is for fine slips. This is because the coarse grain pack has many grain-to-grain contacts, producing a flow characteristic not unlike that of concrete. In both cases, the large particles dominate.

Viscometers

Cup and cone viscometers are not suitable for coarse slips. One can use rotor-type viscometers for coarse slips. Unless the grain is too coarse, in which case viscosity measurements would not be especially meaningful.

Milling time

The coarse fraction is not milled down to finer sizes. Milling, instead, is done to disperse the fine fraction that will act as a bond in the sintered structure. A milling function of the coarse fraction is to act as a grinding media assisting the deagglomeration of the fines. Therefore, milling time is shorter than that used in fine slips. The size of the coarse fraction is controlled by batching rather than by milling. A coarse slip may be milled for only a few hours to disperse the fines.

Volume Percent Solids

It is a common practice to batch coarse slips to a much higher solids content. The coarse-grain pack determines the overall volume of the slip. Intermediate and fine sizes fill the interstices. Solids up to 85% are common, where 50-55% are typical for fine slips. Techniques for obtaining these high percentages will be discussed in the section on forming in Chapter VI.

Surfactant Concentration

The grain pack dominates slip rheology. However, the fine sizes have to be deflocculated. Since the amount of coarse material is large, less surfactant is needed in proportion to the amount of fines. Surfactants are ineffective on coarse particles as they have much less surface in proportion to their mass, gravity dominates.

Binders

Often, one uses coarse materials for refractory applications where cost is an important factor. Normally, one would not use acrylic binders, but methyl cellulose, waxes, gums, and PVA are often used. Binders do not have to impart a high strength to the green body as it will not be green machined and ordinarily is not formed into a delicate shape.

pH/Zeta Potential

Since gravity controls the rheology, these measurements are not generally made. Some knowledge of how the fines behave is relevant. One can ascertain this by viscosity measurements.

Settling

Settling is an important property for coarse slips. The large particles settle out very quickly and result in segregation with the coarse grain on the bottom and fine grain on the top. During sintering, the top will shrink more than the bottom, resulting in distortions and fractures. Even before that, the slip will segregate in the drying process, requiring remixing. It is difficult to prevent segregation since it often happens in dry grain storage. In a slip, one can slow the settling process by increasing the viscosity, often with a thixotropic additive such as Hectorite, or a soluble polymer. Although not a coarse material, latex paint is an example of the use of these additives. Slip rolling is discussed a little later in this chapter. Rolling prevents settling of the coarse fraction during storage.

Check List, Coarse Particle Slips

In dealing with course particle slips, one needs to pay attention to factors as listed below.

- Particle size measurements
    - Screen selection
    - Contamination
    - Sieving
- Viscosity
    - Dilatant
- Milling
- Volume % solids
- Binders
- Settling, can be a serious problem

5.0 SOLIDS RECOVERY

Occasionally, it is necessary to recover the solids from a slip especially when the water contains dissolved impurities that have to be removed or when the slip has been leached, often with an acid. There are a few commonly used ways to recover the solids.

Sedimentation

When a slip is placed in a vessel, it will settle out. The settling rate is determined by Stokes' law. High density and coarse particles settle faster. Liquid viscosity and high solids slow the settling rate. It is not practical to recover solids of a sub micrometer particle size unless they are flocced. Floccing will result in a soft cake, which makes it easy to remove. Fine grained dispersed slips will settle out as hard as a rock making it difficult to remove them from the vessel. Once the slip settles out, the supernatant liquid can be decanted or siphoned off. Please do not suck on the siphon tube to start the siphon. Fill the siphon tube with water, and with a finger at the bottom of the siphon tube, stick the tube into the vessel.

Settling will segregate the particles with coarse and higher specific gravity particles at the bottom. The very fine material and solid organic contamination will be on the top. When dried in this condition, the top slime will form hard agglomerates that are difficult to remix. Often, after settling, the material is re-blunged in clean water or in dilute acid for a second wash, and repeated as often as necessary. Following this, the material is dried or made into a slip for further processing. It will segregate unless it is remixed after drying.

Filtering

One can use a variety of filters to recover the solids. In the laboratory, only some few techniques are ordinarily available. The Buchner funnel, shown in Figure 4.21, is especially useful. The funnel is porcelain with a perforated plate that holds the filter paper. A vacuum is drawn on the lower portion creating a pressure differential that increases the filtering rate. Often, an aspirator is used to draw the vacuum. There is a technique for using a Buchner funnel (filter) to produce the best results and avoid having the slip disappear down the drain.

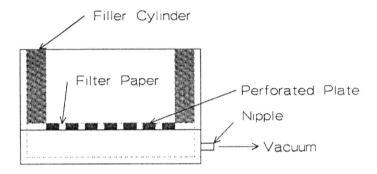

Figure 4.21: Buchner Funnel. The funnel is used to recover solids by filtration.

The fineness of the filter paper determines particle retention. Use a hard paper. Often, one uses a slightly coarser paper because it filters faster. When a thin filter cake forms, the cake itself will act as the filter. With the cake in place, one can recover material that initially passed through the filter. Placing the filter paper on the perforated porcelain plate is tricky. One needs to center the filter paper, smooth it out, and then wet it with water. Following this, one then pulls a vacuum on the funnel to seat the paper without wrinkles. Wrinkles are the main cause for losing the slip. The funnel is constructed with the perforations placed a distance from the edge, resulting in a dead volume around the perimeter. This is of no concern unless the material is to be washed on the filter; which means the edge will not be washed and will contaminate the batch. There are two ways to avoid this: re-blunge the material and repeat filtering or cover the dead space around the edge with an insert as shown in Figure 4.21. A vacuum is usually obtained with an aspirator, however a vacuum pump can be used. The pump will pick up water vapor and slip unless it has adequate traps. Pour the slip onto the center of the filter. Pouring the slip down the edge can dislodge the filter paper and cause the slip to be washed down the drain. When washing on the filter, the cake will crack if left too long where it dries in the funnel. The crack will act as channels for the wash water. If this happens, re-blunge the material and filter again. The best way to remove the filter cake from the funnel is to tip the funnel up so that all of the filtrate drains out of the base. One then has to clean the nipple, turn the funnel upside down over a pan, and place the lips on the nipple and blow. The cake and paper will then drop out neatly. If the cake is to be dried at this point, it is much better to leave the paper on because removing the paper when wet can cause it to tear, leaving pieces in the batch. To clean the Buchner funnel, back flush it through the nipple while still wet and tip the funnel around to flush all locations in the base. When left to dry without cleaning, the solid will cake and be hard to remove.

Another filtering technique is to use a pressure filter. This device is especially useful for filtering very fine slips that are viscous. A lab pressure filter (Millipore) is shown in Figure 4.22. The filter has a slotted base plate, a shell, and a top plate with gaskets to make the seals. A filter medium is chosen and placed on the base. Pour in the slip and pressurize the apparatus to about 60 psi. Always use a pressure relief valve that limits the pressure below the rated capacity. Turn up the pressure to just above 60 psi to assure that the valve is operating correctly. Pressurized gases are very dangerous. Dried ceramic slips on the valve port will render it useless. Also, routinely check the pressure gauge to double check safety.

Figure 4.22: Pressure Filter Apparatus. The filter is used to remove extraneous material. (Courtesy of Millipore).

Centrifuge

Fine, particle-sized slips can be difficult to recover by other methods, and centrifuging can come in handy. There are three main types of centrifuges: solid bowl, overflow, and filtering. The solid bowl packs the solid particles against the sides or base by increasing the gravitational force and causing an increase in the settling rate. The bowl is a cylinder that rotates around its axis. After centrifuging, the liquid on top is decanted. For small quantities of material, a lab centrifuge with polymer tubes is useful. Lab centrifuges have a rotor with holes into which one inserts tubes, like test tubes. One commonly uses polymer tubes such as polyethylene as glass tubes break. Pour the slip into these tubes and spin them until the solids sediment out. These centrifuges are available up to high gravitational forces, but are limited to the amount of material they can handle. Even then, when the particles are very fine, this is a practical lab approach. At lower speeds and if the slip is flocculated, solid bowl centrifuges can recover materials in the sub micrometer range. A problem is that the cake is extremely dilatant and exceptionally hard to remove. Filtering centrifuges have a filter medium around the bowl that separates the solids from the liquid. Problems are similar to those with filters. Decanting centrifuges are essentially related to the solid bowel centrifuge, but the effluent is a liquid containing the finer particles. The decanting centrifuge is not so much for the recovery of solids as for classification. An example is shown in Figure 4.23.

One technique is to circulate the slip through the centrifuge and an ultrasonic disperser, with the effluent comprising the fraction recovered.

Figure 4.23: Decanting Centrifuge. Slip enters into the bottom by a peristaltic pump. Fines are taken off up the shell and can be recirculated to remove additional coarser particles. (Courtesy of IEC)

Casting

This is an old potter's method for recovering solids. How simple! Just pour the slip out onto a plaster plate and have the plaster remove the excess water. When there is a concern that the plaster may contaminate the ceramic powder, place a flat disc of filter paper on top of the plaster to act as a separating plane.

Check List, Solids Recovery

For recovering solids, observe the following methods:
- Sedimentation
- Filtering
  - Buchner technique
  - Pressure filter technique
- Centrifuge
- Casting on plaster

6.0 SLIP CONDITIONING/STORAGE

Since there are differences between fine and coarse grained slips, we will address them separately.

Fine Particle Sized Slips

Surface area increases as the inverse square of the particle size. Fine slips are in more rapport with the liquid, than coarse slips that are in rapport with gravity. It can take much longer for fine slips to equilibrate with the liquid, perhaps as long as 24 hours. During this interval, the viscosity is changing (increasing). An industrial practice is to incrementally add slip allocates to a master batch that has had time to equilibrate, resulting in a consistent slip for processing. Viscosity measurements, when leveling out, will suggest when the slip is stabilized. The author has heard that in ancient China, the master lays down a bed of clay for the next generation to make beautiful ceramics.

Equilibration

Particles in the slip are undergoing adsorption/desorption reactions with electrolytes and surfactants in the liquid. These reactions can take time to reach equilibrium, perhaps as long as 24 hours. During this interval, the slip is altering its viscosity and casting characteristics. Fresh out of the mill, the slip will be warm - or even hot - and can cook organic additives. Reaction rates are very temperature dependant, and equilibrium at an elevated temperature may not be equilibrium at room temperature. If this slip needs time to equilibrate, it is a good idea to age it before using it. Not all slips have this problem. The slip is stored during the equilibration period, and this is the subject of the following section.

Filtering

Filtering was discussed in the preceding section on solids recovery. However, filtering the slip to remove irregularities is a different subject. Fine slips will not pass through a filter in the usual manner since they are too viscous. It is necessary to pressure filter these slips in the apparatus previously shown in Figure 4.22, only here the filter is a cloth mesh such as that shown in Figure 4.24.

This filter cloth has an opening of 20 μm, so anything finer than that can pass right through. Meshes as fine as 10 μm are available, but slow. Filtering removes clumps of binders, agglomerates, and extraneous materials. Figure 4.25 shows some debris recovered from what was thought to be a clean slip.

Figure 4.24: Filter Cloth with 20 μm Opening. The material is regular and clear for material to pass through.

Figure 4.25: Debris Recovered from a Slip by Filtering. Even a high purity slip can contain extraneous material.

There are three materials in the photo: paper towel fibers, wood sticks, and agglomerates. Fibers from paper towels are endemic in the lab. Try the following experiment. Put on a pair of disposable plastic gloves, wash your hands, and dry with a paper towel. Now, rinse the gloves with a squirt bottle containing acetone (inflammable) onto a watch glass and let it dry. There will be lots of fibers on the watch glass. Electronics and pharmaceutical industries are aware of these contaminating problems. People are also a major source of contamination, and this is why we see pictures of people in these industrial labs wearing lots of protective clothing. This is done not to protect the people, but to protect the materials. Pressure filtering will remove the larger (20 μm) contaminates. Careful handling might cope with the smaller contaminating particles if they were not put in by the powder supplier.

Pressure filtering is not difficult. After the filter is set up, one pours in the slip, closes the apparatus, and pressurizes it to about 60 psi. Cleaning up is the problem. After each run, one dismantles the apparatus and thoroughly cleans it. A buildup of dried slip is a source of hard agglomerates and can cause the apparatus to malfunction. The filter cloth cannot be cleaned effectively. It can, however, be used again with the same slip composition if handled and stored carefully.

De-airing

Bubbles are entrained in the slip during milling. If left in, they will result in voids in an article that is slip cast. Small bubbles are difficult to remove except by vacuum de-airing. The apparatus for de-airing is shown in Figure 4.26. The apparatus consists of a vacuum chamber, a polymer liner, and a view port. Quick disconnects and a rubber hose connects the chamber to the vacuum system. There is also a stirrer to mix the slip during de-airing. A view port is necessary for observing the slip during the de-airing process. Both the view port and lid have gaskets for sealing. There is no need for clamping as the vacuum exerts enough force to make the seal. The polymer liner is there to help clean up. The vacuum system includes the disconnects, hoses, valving, vacuum gauge, a desiccant cartridge, air filter, and a fore pump. A schematic of this system is shown in Figure 4.27.

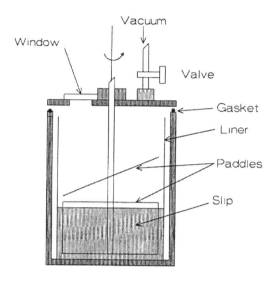

Figure 4.26: Vacuum De-airing Apparatus. Valving provides for de-airing of slips and de-airing of dry powders.

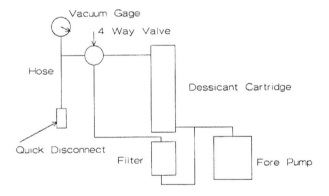

Figure 4.27: Vacuum System, Schematic. A desiccant or cold trap is essential.

The system is mounted on a roll-around cart with locking casters. It has a variety of functions. For example, the filter protects the vacuum pump when evacuating isopress tooling containing powders. Desiccants that change color when saturated are available and one can reactivate such cartridges in an oven. A cold trap is an option to desiccants. Liquid nitrogen (LN_2) is preferred, but one can use mechanical cold traps if LN_2 is not readily available. The fore pump must be able to attain a vacuum of at least 2300Pa (17 mm Hg, 29" Hg), for reasons that will be explained.

To operate the de-airing function one is to pour the slip into the liner no more than a one third level, close the vacuum container, start the slow paddle rotation, and pull the vacuum. When the pressure reaches about 2300 Pa, the slip will froth up to three times its volume. This is why the chamber has the paddle above the original fill and the fill is limited to a one third level. There will be a mess if the liner overflows, limiting the quantity of slip that can be de-aired in one fill. The froth will then collapse under its own weight, and continue to bubble for a little while. The slip is now de-aired.

When the binder in the slip forms a leathery skin due to drying in the vacuum, the paddle will stretch the bubble walls and rupture them. This type of slip is difficult to de-air without stirring. An alginate binder system forms such a skin.

After de-airing, one releases the vacuum and lifts out the liner. One should immediately clean the liner as wet slip is a lot easier to remove. Dried slips can be rock hard depending on the binder.

With the operational procedures out of the way, the physics of the process can be discussed. A liquid boils when the pressure above the liquid surface is equal to the vapor pressure of the liquid. Water boils at 100 °C at 101 kPa (760 mm Hg, one atmosphere), or at 2.34 kPa (17.5 mm Hg) at 20 °C. In each case, the vapor pressure of the liquid is equal to the pressure over its surface. Entrained air bubbles act as nuclei for bubbles of water vapor, which greatly increases the bubble volume. This increases the buoyancy of the bubble so the slip froths up with the water vapor constituting the bulk of the gas volume, and the air just going along for the ride. The froth collapses under its own weight, releasing the water vapor and the entrained air pumped out of the system.

Non-aqueous slips can also be vacuum de-aired, but there are complications. Often, these slips contain two or more organic liquids, each of which has its own vapor pressure. The liquid with the highest vapor pressure is selectively removed. Also, some plastics are soluble in organic solvents, some plastics will solvent craze, and other plastics will adsorb solvent vapors and swell. These problems are solvable with the judicious selection of construction materials for the apparatus. When using solvents, the preferred cold trap is LN_2.

Magnetic Filtering

When one processes a material through equipment that can add ferrous contamination, one then passes this material through a magnetic filter. Dry filters are often made from permanent magnates and the mix is poured through a baffled filter. One can only remove ferromagnetic particles and large particles this way; however, this filtering process can be good enough for some applications. Figure 4.28 depicts a slip filter using electromagnets.

Figure 4.28: Magnetic Filter for Slips. Magnetic field strength depends on extraneous material to be removed. (Courtesy of Eriez Magnetics)

Field strength selection is dependent on the removal of ferro- or para-magnetic contaminates and on the particle size of the magnetic particle. High field strengths are for weakly magnetic and/or small particles. One can best separate ferromagnetic large particles; it is more expensive to facilitate better separations.

Slip Storage

Often, the slip is stored for later use or for equilibration. Seal the slip storage container and slowly rotate it to prevent settling of the particles. One can use a modified mill rack provided one lowers the speed to about 1-5 rpm. If the speed is too high, the slip will cascade and reintroduce bubbles. For lab use, a one-liter glass jar is a convenient size. There are three problems with glass storage jars: the jar lip is not flat, the slip reacts with the lid liner, and the jar is fragile. Jars with a bump on the lip prevent an air tight seal. This causes the slip to dry around the lip forming moderately hard agglomerates that drop into the slip when the jar is opened. Lap the jar lip with SiC abrasive, coated or loose grain. Lapping the lip flat and smooth is a little tricky. Friction between the lap and the jar produces a torque that will tip the jar when the jar is held up from the lap surface. One can reduce the torque by holding the jar at a low level. When the jar is held too low, the abrasive when in contact with your fingers will remove skin from your fingers and possibly cause them to bleed. Surprisingly, the finger tips are insensitive to this abrasion. The combination of the cold water and vibration probably accounts for this insensitivity. Please accept my testimonial on this.

Lid liners sometimes react with the slip or are softened by it. A good practice is to cut a rubber gasket to fit into the lid. Neoprene is a good choice. This gasket will not react with the slip and will make a good seal with the lapped jar lip, which will prevent drying and the formation of hard agglomerates.

Wrapping the sides of the jar with duct tape will inhibit breakage, or at least hold the fragments together if the jar is broken. This will also increase friction on the mill rack so that the jar rolls smoothly. Tape wrapping also identifies the jar as modified for exclusive slip rolling use.

After taking a slip aliquot from the jar, one should transfer the slip to a clean jar. This transfer prevents agglomerates from getting into the slip from the gasket or lid.

Check List, Fine Particle slips

One should observe the following factors when dealing with fine particle slips.
- Slip equilibration
- Pressure filtering
- De-airing apparatus
- De-airing procedure
- Magnetic filtering
- Slip storage apparatus
- Slip storage procedure

Coarse Particle Slips

Often, one augments the coarse fraction of the slip by additions of coarse grain. There are advantages to doing this such as: increased green density, lower firing shrinkage, improved resistance to thermal shock, and improved resistance to erosion. Since one makes most coarse slips by mixing, this discussion is on casting slips.

Particle Packing

Figure 4.29 shows the particle packing of three constituents: coarse, medium, and fine fractions. Each is a screen size cut with a top size and all of the finer material that passes through that screen. For example, 24F (through 24 mesh and all of the finer sizes) is a typical cut. The ratio of the top size diameters for the three cuts is best determined experimentally where the maximum packing density is obtained. Dry packing is measured by the weight of material that will fill a cylindrical container with the top of the fill scraped off level. One can either tap or vibrate the table holding the cylinder. This is tricky as the grain will segregate with the fines moving toward the bottom. In this case, one compromises the packing of the various sizes. Since the concern here is with slips, a different technique is called for. As a generalization, the maximum diameter of the constituents is at a ratio of about 10:1-7:1 for each split. Figure 4.29 shows the approximate amount for each.

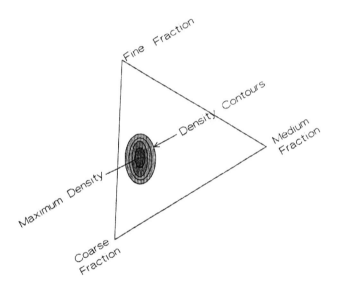

Figure 4.29: Particle Packing Diagram. Maximum density is in the approximate location shown for three size fractions.

Taking pure alumina as an example, the fine portion will be a Bayer alumina with choices of diameters from 0.6 μm to 4.0 μm. The fine portion has to sinter to make the bond, so it has to be fine. With other materials, low cost fines may not be available and one may have to use a -325-mesh fraction. Micrometer-sized particles are available for a variety of ceramics, but they are expensive.

There are theories for optimum particle packing for both precisely defined sizes and for distributions. These are not directly applicable as there are other variables such as particle shape. For example, in one case a higher density for a SiC refractory was obtained using mulled grain. With the sharp corners removed, the particles packed to a higher density. The practical way to optimize dense packing is by experimentation, but mixing to attain that is an important variable.

For coarse slips, it is common practice to increase the amount of coarse grain to increase the solid content. Since coarse particles are impervious and dense, when they are substituted for a volume of slip, there is an increase in density and a net increase in the interstitial water. The technique for doing this is by rolling the slip, as introduced in the following section.

Rolling the slip

It is feasible to increase the solid content to 85% by rolling the slip and adding coarse grain in increments. There are two concurrent processes: coarse solid particles replace fine agglomerates and the packing becomes denser as the slip is rolled. As the slip is rolled in a slowly rotating mill, it becomes less viscous. This is because the particles pack more efficiently while returning the water to the interstices. One can add a little more coarse grain to the batch and keep repeating the process until the slip thickens.

One can postulate an explanation for what is occurring. In the rolling process, the slip is subjected to shear. In shear, the particles are sliding over one another and are being rotated. Velocity gradients in the liquid rotate the particles as the fluid velocity on one side of a grain is greater than that on the other. There are short range domains where the packing density is increased. Figure 4.30 illustrates this idea.

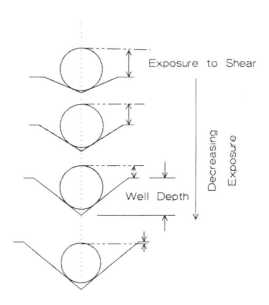

Figure 4.30: Packing Energy Wells. Particles will work themselves into deeper wells when the shear rate is appropriate.

Particles rest in energy wells in the grain pack. When the shear vector is high enough, the particle can be displaced. By chance, it will drop into another energy well that may be deeper than the former one. By so doing, it is more tightly packed and harder to displace. Shear gradient, viscosity, and particle size decide the energy level available in the shear field. One can increase shear by increasing the mill speed. However, too high a speed will displace all of the particles and may warrant the grain pack to be dug out of the mill with small hand tools. There is an optimum short range of mill speeds for best packing. One can determine this with experimentation.

High solid slips are dilatant, at least in part, while the densely packed domains retain their identity in the slip. Think of slush balls in mud. When these slips are poured from one bucket to another, they writhe. It would not do so without short range ordering.

Storage of coarse-grained slips is done by rolling, but unlike their fine grained counterparts, there is a top speed above which the slip will set up. Settling is a major problem for coarse-grained slips, not so much in storing as in slip casting. One can reduce settling by either increasing the slip viscosity with polymers or inorganic colloids such as hectorite or by speeding the casting rate. Pressure casting is a very good idea for these slips.

Check List, Slip Conditioning and Storage

Observe the following factors for slip conditioning and storage:
- Adsorption/desorption equilibration
- Filtering
- De-airing
- Slip sorage
- Particle packing
- Slip rolling and solids content increase
- Rolling in storage
- Settling out of coarse particles
  - Increase viscosity
  - Pressure casting

REFERENCES

1.   Fred F. Lange, "Powder Processing Science and Technology for Increased Reliability." J. Am. Ceram. Soc. **72** [1], 3-15, 1989.
2.   Greenleaf, J. Personal Communication.
3.   Anne Bagley Hardy, Wendel E. Rhine, and H. Kent Bowen, "Preparation of Spherical, Submicrometer Oxide Particles by Hydrolysis of Emulsified Alkoxide Droplets." *J. Am. Ceram. Soc.* **76** [1], 97-104, 1993.
4.   K. S. Venkataraman, "Predicting the Size Distributions of Fine Powders During Comminution," *Advanced Ceramic Materials*, **3** [5], 1988.

5

Mixing Coarse Grained Materials

1.0 MIXING CONSIDERATIONS

Whenever one uses a coarse mix for extrusion, pressing, or ramming, it is not always made up as a slip. Usually, it is mixed dry or semi-dry in a mixer or blender. Perhaps, the primary distinction between mixing equipment is the intensity of mixing. Current practice is toward higher intensity mixing. The other requirements include uniformly mixing all of the batch, recovery of the mix, and cleanup.

The impeller is generally thought to do the mixing by impact with the grain. While there is some of this, autogenous impact between the particles themselves may be the primary deagglomerating process in the fines. The impeller establishes shear in the mix which invigorates the grain. In turn, the grain particles impinge upon each other and do most of the work. It can be looked at as a miniature sand mill where the coarse particles themselves are the media deagglomerating the fines. The purpose of mixing is to generate a uniform distribution of particles of different size, shape, and density. One should review the extensive literature available on this subject to consider the possible variables. One can start by reading papers on the grading of aggregates by Furnas and Anderegg. These papers and other literature address the ratios of the volume fraction of each constituent for obtaining the highest density, which is usually the objective unless the ceramic is for a permeable application. Figure 4.29 shows the general location of highest density for three sizes of particles. Design your experiment around this region in the diagram. Dry mixing and wet mixing will be discussed.

2.0 DRY MIXING AND EQUIPMENT

It is easier to handle ceramic mixes when they are dry or damp as wet mixes stick to everything. One uses dry mixes for the following applications: mixing ingredients for calcination to form a different phase such as with a ceramic pigment, pressing to form a shape with refractories being a common application, and ramming where the shape is unusual such as in a foundry induction furnace lining or for a long coarse-grained tube or rod. Large SiC resistors and long tubes can be end-rammed tamped.

Making a uniform mix is difficult in at least two ways. Coarse or dense grains will easily segregate into coarse, fine, light, or heavy fractions. It is very difficult to minimize these tendencies. Secondly, it is difficult to determine if segregations are present in the mix. Extreme cases are visually apparent, but too often one cannot tell by casual observation. This is further discussed in a following section on mix uniformity.

There are five types of dry mixers: high intensity, mullers, sigma blade, V blenders, and ribbon blenders.

High Intensity Mixers

Perhaps best known is the Eirich type of mixer. There are a few variations, but the one with a high-speed, vertical-shaft impeller and plow blades is most common. The mixer has a rotating pan tipped up from the horizontal. The impeller is parallel to the shaft axis and has a head with impeller blades. These blades can be cemented carbide for wear resistance. There is also a plow blade that is stationary, helping to distribute the mix. The shaft housing and the plow support are fastened to the lid. This mixing action is unique in regards to distribution of the material during mixing. This is illustrated in Figure 5.1.

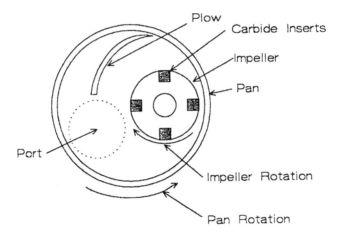

Figure 5.1: High Intensity Mixer Drawing. The flow of material is back into the mixing head.

In this situation, the mixing process is thorough. The impeller rotates at a high speed, the plow moves the mix back into the center of the drum, and the drum counter rotation eliminates any dead spots where the mix is stationary. Another advantage is that the seals and bearings are up out of the mix, thus extending their productive use.

Figure 5.2 depicts a commercially available high intensity mixer (Courtesy Eirich). The mixer is tilted to cause the mix to cascade back into the mixing impellers. This mixer has two speeds by changing pulley combinations. A continuously variable speed control would be preferable.

As was stated earlier, the coarse particles themselves do much of the fines deagglomeration. Visualize what happens in the shear zone as mixing occurs. A schematic is illustrated in Figure 5.3. Particle impingement in a shear field deagglomerates the fines.

Figure 5.2: High Intensity Mixer. Speed variation by the choice of pulleys.

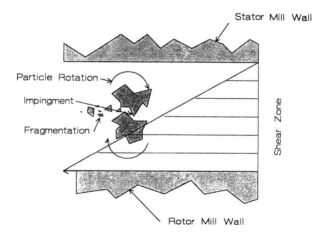

Figure 5.3: Deagglomeration by Shear

A velocity gradient persists across the shear zone, causing the agglomerated particles to rotate. In Figure 5.3, the two large particles are rotating in the same direction, but their surfaces are impinging in opposite directions with the contact velocity doubled. Impact between particles can result in grinding. Large particles will have more energy and will be rotating at a higher velocity than the fine particles. Consequently, the fines are ground up by the coarse aggregate. Then, whenever the mix contains coarse material, the fines will be ground quicker than if the larger particles were not present. This all makes the grinding time a function of particle size. Coarse particles act as grinding media in the mixer. A similar mechanism also occurs in fluid slips.[1]

Mullers

Mullers have been used successfully for a very long time. They can be used for grinding and mixing depending on the hardness of the material and the weight of the muller wheels. A softer material like wheat is ground to flour with mullers. A harder material like SiC is mulled to knock off the sharp edges.

Muller wheels can be massive. For ceramics applications, they are usually iron as the weight helps to mix and grind the batch. In agriculture, the wheels traditionally are stone hewed out of massive blocks. When iron contamination is not tolerable, one can also use stone wheels for ceramic applications. The pan rotates counter to the two wheels, with the plow shepherding the batch into the grinding zone. The wheels also rotate on their own axis resulting in a shearing zone between the pan and the wheel surface. This wheel assembly rotates about the pan axis. Since the outermost part of the muller wheel is set out farther than the part nearer the pan axis, it moves faster and the mix is smeared as it is mulled. It is the shearing effect, primarily, which does the deagglomeration of the fines in the batch. It is not so much the purpose of the muller to grind up the coarse grog here, however the muller deagglomerates the fines and homogenizes the mix. Figure 5.4 depicts a sketch of a muller.

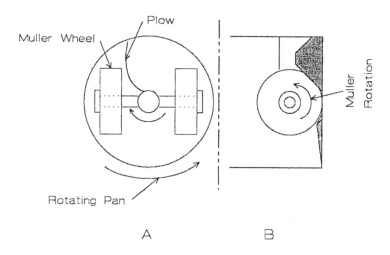

Figure 5.4: Muller Mechanism Drawing. Shear between the wheels and pan base deagglomerates the particles.

Mullers have a bottom port that can be opened to discharge the mix, with the plow forcing the mix into the port. Bearings are out of the abrasive mix, making the equipment durable. Clean up is easy as the pan can be hosed down, with the water exiting out of the port. Mullers are useful and will probably be used at least for a few more millennia. A commercial lab-sized muller is shown in Figure 5.5. The muller wheels and side can be removed for easy cleaning.

Figure 5.5: Laboratory Muller (courtesy of Simpson Mix-Muller)

Sigma Blade Mixers

This type of mixer is a trough with two sigma-shaped arms that rotate to blend the mix. One can use either pug or sigma blades, but they are not interchangeable without dismantling the mixer. Sigma, named after the Greek letter, is shaped like the letter S. Two blades intertwine while rotating, resulting in a moderate shearing intensity and homogenization. The trough tips to dump the batch providing for easy clean up. The seals and bearings are in the batch during mixing, requiring maintenance. In this mixer, the shearing effect is not intense enough for fine-grained batches, but one can satisfactorily mix relatively coarse mixes.

V Blenders

This equipment consists of two cylindrical sections joined together to make a V shape as shown in Figure 5.6.

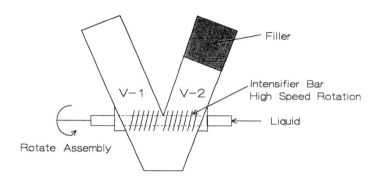

Figure 5.6: V Blender. Unequal volumes on the two sides help to cross blend. The intensifier bar aids in deagglomeration and is used to add binder solutions.

The entire assembly revolves around the axis, tumbling the mix repeatedly from the base to the top of the cylinders. When the capacity of the two cylinders is different, there is transfer of material from one side to the other, as suggested in Figure 5.6. There is the option of adding an intensifier bar that also rotates on the axis but at a much higher speed. The bar increases the shearing effect in the mixer; it also has a central channel for adding a liquid, such as a binder solution. This is a very useful feature, as the batch can be blended dry before adding the binder. Being able to observe the mix consistency during addition of the binder solution takes some of the guess work out of just how much to add to obtain the desired consistency. Too little binder solution and it will not press properly. Too much binder solution and the mix will coat the sides creating a compromised batch and a cleanup problem.

Ribbon Blender

Like the sigma blade mixer, ribbon blenders are horizontal troughs with a small number (about three) of helical blades that rotate around the trough axis. Like the sigma blade, the trough tips to dump the batch. Mixing is gentle, making ribbon blenders useful for premixing batches or for calcining mixes.

How well is the batch mixed? There is a quick way to tell when the materials are of a different color. Pour a little of the mix onto a piece of white paper and run one's finger across it. Colored streaks reveal that the agglomerates have not been broken down. At most, this is undesirable and one should either mix longer, mix at a higher speed, or mix with a higher intensity mixer.

3.0　WET MIXING AND EQUIPMENT

Wet mixing is used where the material has to flow during the forming process. Dry mixes have very little flow. Wet mixes are plastic similar to modeling clay. Examples are extruded parts such as coarse-grained tubes and deep-crucible shapes where the mix has to flow up the sides of the mold to fill the cavity, a mud pressing process.

Two types of wet mixers will be discussed: pug mills and extruders. Both have a horizontal axis mixing arm with blades in the pug mill and a screw in the extruder. In both cases, the mix is forced down the length of the mixer as the mixing arm rotates. Consequently, one batches the mixes in a separate apparatus to feed a homogeneous mix into the mixer. In the polymer literature, there is extensive material on the topic of extrusion. The difference being that the volume per cent of solids is much lower and the molten polymer dominates the rheology. Additionally, the polymer remains in the finished article while with ceramics it is burned out. Still one might want to refer to these references.[1,2]

Pug Mills

The mill is a trough with a horizontal mixing arm with angled blades that shear the mix and move it from one end to the other. Since the mix is viscous, the arm is massive and driven by a high horsepower gear reduction drive to withstand the stresses. The rotational speed and viscosity of the mix govern the mixing intensity. Speed is limited by the stresses imposed on the apparatus, so most of the mixing is a consequence of the high viscosity. One uses pug mills to prepare a mix for extrusion as preliminary blending is often necessary.

Extruders

Auger Extruders

An auger extruder is similar to a meat grinder for making hamburger meat in a butcher's shop. It has an entry port feeding the mix down into the auger where it is mixed and extruded out of a die. A laboratory auger extruder is shown in Figure 5.7.

Wear is intensive and the barrels often have replaceable liners. Both augers and blades can be coated with a wear-resistant material. Figure 5.7 depicts an extruder with a variety of features. One feeds the mix from the right side into the hopper where an auger forces it into a rotating cutter. The function of the cutter is to shred the mix and to remove the entrained air in a vacuum. It is necessary to feed fast enough to maintain the vacuum in the machine. After shredding and deairing, the mix is dropped into a second auger that mixes and compresses the mix into a solid mass. The mass is then forced through a die with a size and shape as selected. The size of the die has to be appropriate with the feed rate, since if it is too small the machine will overload. A useful feature of this extruder is that the working parts are fastened to the base with toggle clamps for quick change and cleanup between mix compositions.

Figure 5.7: Laboratory Extruder. The extruder has a port for adding material, a chopper in a vacuum chamber, and a die. (Courtesy of Fate International)

Wire/Knife Cutting. The extrusion is usually wire cut to length. Wear on the wire is severe and breakage can be a serious problem as it shuts the process down. This can be avoided by slowly feeding the wire from a spool, through a tensioning device, and onto a take-up spool. In this fashion, fresh wire is continuously supplied to the cutter.

Mixing efficiency. Not all of the mix receives the same amount of shear in an extruder. Material next to the wall of the barrel is not mixed as intensely as material in the middle of the flight. There can even be dead areas where the mix is stationary. Since mixing is not uniform, the extruded piece is not uniform in green density or in breaking up of agglomerates. There are many extrusion problems almost all of which are attributable to poor mixing. This

subject will be expanded in a later section in this chapter. The following can be done in case of an extrusion:
- send the extrusion back for a second and/or third pass,
- premix in another apparatus,
- increase the shearing effect by using a multi-orifice die with small openings for preliminary mixing, or
- change the viscosity of the mix.

Ram Extruders

A second type of extruder uses a hydraulic cylinder rather than an auger to force the mix through the die. One supplies the mix to the ram extruder as a premixed cylinder that drops into place in the barrel, as ram extruders do not mix the material. This will be discussed in greater detail in the chapter on forming. One can set up ram extruders with two barrels that alternately rotate in place, making the process almost continuous.

Hydraulic ram extruders supply material at a constant rate. Cutters can be time controlled. Another way to control the cut is with a fiber-optic sensor and an actuator, such as a solenoid connected to a wire cutter. Tolerance on the length of cut can be as precise as 0.005 inches.

Combinations of equipment

These combinations have been alluded to in the previous discussion. A mixer, pug mill, and extruder can be coupled into a lab process line. This is not at all uncommon and is often necessary. Most lab work, however, is not
that structured, and mixes are often carried from one piece of equipment to another. Figure 5.8 is a schematic of a combination process line.

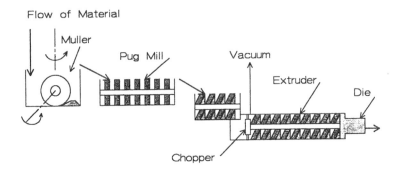

Figure 5.8: Mixing Process Line. Mixing is often done in three steps: mulling, pug milling, and extruding with deairing.

The material is batched in the muller, then transferred to the pug mill. The function of the muller is to distribute the mix ingredients uniformly on an overall scale and deagglomerate the fines. All of the ingredients are batched in the muller and mixed as a batch. This function can also be done in a pug mill if the port emptying the mill is closed. In this case, there is very little linear transfer of material. The function of the pug mill is to homogenize the material by shear and to provide a more uniform batch to the extruder. The extruder sketch shown has two compartments. The first is to introduce the material into the extruder and force it into the chopper. A vacuum is pulled in the chopper chamber to remove entrained air. (Chopping exposes the air pockets to the vacuum.) Then, the chopped mix is dropped into the extruder auger, where it is intensively sheared and mixed. Auger flights move the material to the extrusion die under pressure; this compacts the mix and forms it into the die configuration. It is not uncommon to use only the muller and extruder for processing. When more mixing is needed, the pug mill can be put into the line.

Special Purpose Mixers

Many other type of mixers are available other than those discussed. One of these is a closed, vacuum-tight chamber with a jacket for either heating or cooling. Figure 5.9 is a lab-sized mixer of this type.

Figure 5.9: Special Purpose Mixer. This mixer provides heat for melting binders and vacuum for deairing. (Courtesy of Ross and sons)

One ceramic application for this mixer is to make up flowable mixes for wax injection. The mixer can be evacuated to remove entrained air and is then heated to melt the wax.

Check List, Mixers

The following is a check list for mixers.
- Amount of Shear: high intensity, mullers, sigma blade, V blenders, ribbon blenders.
- Effect of Coarse Grog on Blending
- Plastic Mixers:
  Pug mills
  Extruders, shaping

- Mixing Lines
- Dry(damp) Mixers:
    High intensity
    Mullers
    Sigma blade
    Ribbon
- Special Purpose

4.0 MIX UNIFORMITY

Coarse-grained mixes will have a large range of particle sizes, from very coarse grog to fines. These will tend to segregate, altering the local mix composition. Almost anything will cause the particles to segregate. They will segregate even when standing in a silo. Proper mixing is to obtain uniformity and to retain it during handling and die filling. The objective of this section is to discuss forces that cause coarse and fine particles to separate during processing and ways to retain that homogeneous mixture.

Segregation of Particles

Gravity

Consider three cases where the coarse-grain pack is mixed with a slip containing the fines: when the coarse-grain pack has a greater volume than the fines, when they are equal, and when the coarse-grain pack has a lesser volume than the fines. Figure 5.10 shows this schematically.

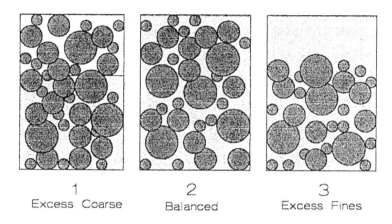

Figure 5.10: Volumetric Grain Packing. Just enough fine material is required to fill the interstices in the grain pack in case 2.

     In illustration 1, the coarse pack fills the volume while the fines do not. When the fill is vibrated, the fines will settle and sift down to the bottom of the cavity and segregate. A body such as this is not uniform and will sinter denser at the base than at the top. The base will be stronger and will have a higher modulus. Coarse packs of the grains will form in every case if the grains are free to move. These coarse packs do not sinter because of the low surface area. In case 2, both the coarse and fines fill the volume and the density is homogeneous; this material will tend to sinter uniformly. In case 3, there are excess fines, and the coarse material will sink to the bottom and segregate. The illustration is also valid for a slip mixture.

     These effects are due to the size distribution in a gravitational field. Particles can also segregate by another mechanism - shear. In a fluid shear field, the coarse particles will preferentially drift to the zone of highest velocity. This is the basis for particle size classifiers such as air or liquid cyclone classifiers. Shear occurs in all mixers. In this case involving coarse grained mixes that are semi-dry, segregation still occurs.

Visualize what is occurring in a mixer. Some sort of impeller is shearing the mix. The coarse grain impinges upon the fine agglomerates and breaks them up. When the mix binder is present in the right quantity including an appropriate level of stickiness, the fines can adhere to the coarse particles in tending to restrain segregation. This will work if the shearing intensity is not so high that it breaks them up again. One way to mix the material is to mix dry at a high intensity to break up the soft agglomerates in the fines and then to slowly add the binder solution at a lower intensity of the impeller. To minimize clumping, the solution can be sprayed into the mix with low intensity mixing. The objective is to pelletize the mix with uniform pellets all of which have the right grain size distribution. The process in all cases for determining uniformity is to sample with progressively smaller sample sizes. When the sample no longer has the batched volume fraction of constituents, one can detect the scale of uniformity. For coarse ceramics, one can apply a series of screen analyses or other particle size analytical techniques when size is relevant. In the case when the density of different fractions is relevant, the mix might be separated by a heavy liquid, with the light fraction floating and the dense fraction sinking. The densest practical heavy liquid is at 3.25 g/ml, making the technique limited in its usefulness. As a postmortem analysis, one can section and polish the material after sintering and observe it microscopically. Chapter 9 describes such an example.

Firing Shrinkage

Constituents in the mix have different firing shrinkages depending upon their particle size. Figure 5.11 schematically illustrates the differences.
Coarse particles shrink little if at all.
Fines shrink up to about 20% linearly. The discrepancy creates a problem in crafting the ceramic. This is illustrated in Figure 5.12 showing schematically the fines that have shrunk away from the coarse-grain pack.

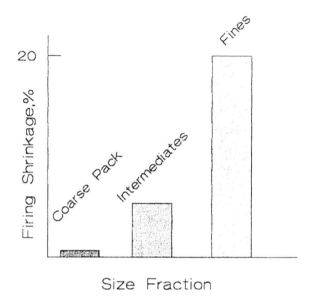

Figure 5.11: Firing Shrinkages. Fine particles shrink more than coarse. The mix has to be balanced to produce a good microstructure.

This, while exaggerated, is a common occurrence that weakens the body. Alleviation of the problem is to add intermediate-sized particles to the formulation to reduce the discrepancy in firing shrinkage. Microstructural analysis will help attain the best formulation. Mechanical properties will detect when something is amiss.

There are four important factors to make a good mix: grain size packing, sintering shrinkage of each fraction, the binder, and the impeller speed. Clues for each are as follows:

Packing. Put the dry mix on a vibrating table and observe what separation occurs or sinter the body and measure the grain size distribution along the length of the piece.

Shrinkage. Add intermediate sizes to reduce the shrinkage of the fines. Also, look at the microstructure of the fired body to see if the fines are shrinking away from the coarse grains.

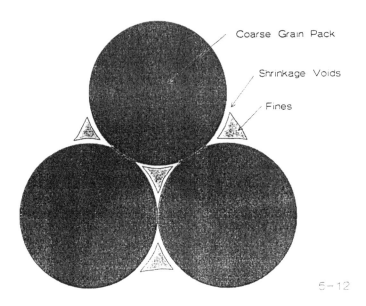

Figure 5.12: Shrinkage Separation of Structure. Fines shrink away from the coarse-grain pack leaving voids.

Binder. There are two things to consider: the amount necessary to hold the body together during handling, and the amount necessary to stick the fines to the coarse particles. One popular way is to grab and squeeze a handful of the mix. If it sticks together without being plastic, then the binding capacity is good. To mix it correctly, it is essential to scrape down the mixer once or twice to get everything back into the bulk of the material. Dry mix first to break up the fines and then add the binder solution slowly at a slower impeller speed.

Impeller Speed. Shear will segregate the mix into fine and coarse fractions. Scrape down the mixer and mix at a slower speed to homogenize the distribution. The angular velocity on the periphery of the impeller is the important criterion for speed.

Check List Uniformity

- Segregation
    - Particle packing
    - Amount of coarse/ medium/ fines
    - Microstructure
    - Pelletize with binder
    - Impeller speed
- Firing Shrinkage
    - Increase intermediate sizes
    - Microstructure

REFERENCES

1.   Alan G. King and Santosh T. Keswani, "Colloidal Mills: Theory and Experiment," *J. Am. Ceram. Soc.*, **77**[3] 769-77 (1994).
2.   Z. Tadmor and I. Klein, Engineering Principles of Plasticating Extrusion. New York: Reinhold, 1970.
3.   Z. Tadmor and C.G. Gogos, Principles of Polymer Processing. New York: John Wileyn & Sons, 1979.

6

Forming

1.0 INTRODUCTION

Forming processes take a mix, slip, or plastic material and form it into a shape. There are many processes available to perform this function. It is usually desirable to have a high green density as this restricts firing shrinkage. In turn, the rejects are reduced and the firing temperature can be lowered. Some of these processes do not result in a high green density and are put at a disadvantage. This chapter describes the following processes: drying, die pressing, other pressing techniques, slip casting, and plastic forming.

2.0 DRYING A SLIP

Either a mix or a part can be dried. The mix is dried before forming to make a press mix; however, a part is dried after forming it to increase the green strength and to prepare it for firing. Drying a part has some different requirements than drying a slip as discussed later in the chapter.

When starting from a slip, drying can be deceptively tricky. The problem is not removing the water. The tricky part is migration of materials in solution to the free surface, crusting of fines on top, coarse particles settling, and agglomeration. One can minimize this problem by spray drying (as compared to pan drying) when the particle size is small. However, one is still dealing with the same problem except on a smaller scale.

Selecting an appropriate method is usually governed by the requirements of the press mix, which can be either damp or dry from using a binder and/or plasticizer. Each drying method will be addressed.

A slip is usually dried to make a press mix, with the drying method selected based on the desired properties of the pressed part. Drying is much more critical for fine-grained, technical ceramics than it is for coarse-grained slips. Coarse materials are permeable with an abundance of fracture sources due to the coarse size. Fine slips form agglomerates that act as fracture sources, degrading the strength. Therefore, drying is an important issue for fine-grained slips. Several drying methods will be discussed.

Pan Drying

This drying method is by far the most widely used in ceramic laboratories since it is so easy to do. One can use stainless pans or glass pans. Place the pan in an oven that has some exchange of air. Since the slip contains binders and other constituents, it is not a good idea to overheat as you could cook the organic components. Sixty degrees Celsius is a reasonable drying temperature. During drying, the coarser particles will sink to the bottom and the *slimes*, which are very fine materials, will migrate to the free surface. In addition, the dissolved materials will also migrate to the free surface, creating a crust. When broken up, the crust will form a distribution of moderately hard agglomerates that upon sintering will shrink more than the rest of the body and create voids. This is not good. Sometimes, the choices are limited and pan drying might have to be the method of choice. High speed dry milling will help to break up these agglomerates, but this is not perfect. Spray drying, stir drying, rotary drying, and freeze drying methods eliminate segregation of the coarse particles and slimes.

Spray Drying

A variety of laboratory-sized spray driers are available. This process has the advantage of producing a free flowing press mix directly. A free flowing mix is very useful because it fills the die cavity easily and uniformly, resulting in a uniformly dense fill. Figure 6.1 depicts a laboratory spray dryer.

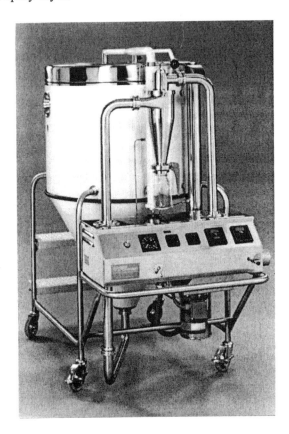

Figure 6.1:   Spray Dryer. Laboratory-sized with controls on air temperature, air flow rate, and slip flow rate. (Courtesy of Niro)

The dryer has several principle parts: a slip pump, a two-fluid spray nozzle, a stainless drying chamber, a conical base, a connecting stainless tubing, a cyclone, and a glass collection chamber. Room air is introduced into the drying chamber by a blower through a heater. Controls include drying air temperature, air flow rate, and a slip feeding rate. On a different dryer than that shown, the following modifications were made: the diameter of the collection chamber was increased to reduce swirling of the granules, and a 12-inch HEPA filter was put on the air intake. The thought was that all of the dust in the air gets into the mix. Clean up is an appreciable task. Thankfully some driers have a body that tips upwards for better access.

To clean, dismantle the dryer by lifting the top and sliding out the nozzle. The glass drying chamber lifts upward and releases the conical base, which is on casters. One can roll out the base for cleaning purposes. The cyclone and collecting chamber are fastened with clamps that are unsnapped and lifted off. This is a lot of work for just one press mix, but what is the alternative? Other lab-sized, spray driers are now available and one should consider them. Cleaning is a very important criterion. Small bench-sized spray driers, because of their size, do not furnish enough time to dry the slip droplets. Therefore, the powder may not have the large agglomerate size needed for flowability and die fill. Buchi (and others) makes a well designed bench spray drier with all of the necessary features. However, to dry the droplets they use a very fine atomizer. This results in a very finely divided press mix that is not especially free flowing.

A nozzle cleaning mechanism is important since there is a tendency for the dry slip to cause clogging problems. There can be two parts: an air jet that blows across the tip, and a needle - coaxial with the jet - that can be forced through the orifice to push out dried slip.

Much of the dried slip adheres to the chamber walls, conical base, and in the tubing. This material is not spherical and most of it can be screened out. Recovery is not very good at 50%; however, one at least gets some good press mix. Figure 6.2 shows the press mix recovered from screening. Particles are semi-spherical with some of them having internal holes. Holes are typical of spray-dried particles and while it is better not to have them, they can be tolerated.

Figure 6.2: Spray Dried Press Mix. Particles are fairly round and show internal cavities. Scale bar 100 μm.

The spray drying process consists of drying the droplets while they are in free fall. They are at least dry enough to avoid sticking to the walls and to each other. To achieve this, the slip should have flow characteristics suitable for atomization. Lab dryers use either two-fluid or vibratory (ultrasonic) atomizers. Commercial spray driers also can use a rotating disc, but lab equipment is too small to adapt to a rotor. Anyway, the slip should be pseudoplastic with a viscosity of about a thousand cps. Droplets must have sufficient residence time to dry. Air temperature, air flow velocity, slip flow rate, volume % solids, the binder system, and the length of the free fall path are important parameters. High air temperature, low air velocity, a low slip flow rate, high solids content, a non-skinning binder system, and a long free fall path are adjustments that will enhance drying. The following are restraints on drying: high air temperature that can volatilize or cook the binder, slow air velocity that lowers the production rate, a low slip flow rate that also cuts down the production rate, a high solids content slip that can cause clogging problems in the nozzle, a binder that forms a skin on the droplet surface, and the free fall path restricted and fixed in a small lab dryer. There will be some experimentation at first to find the optimum drying conditions, which can be different for each material. A good place to start is in the middle of the drying parameters. One can increase the drying process until the dryer's capacity is exceeded or the press mix is not spherical. After achieving this, one can then, back off just a little. For lab applications, there is less interest in the production rate than in obtaining a good press mix, which makes it easier to back off.

After obtaining a spray-dried material, observation will show that there is a range of sizes and shapes in the mix. This mix will not be free flowing. With screening the spray-dried mix, one can remove both the fines, which are the result of satellite droplets and tumbling, and the coarse, which are the result of dried material on the walls and coalescent droplets. With this technique, one can typically get a 50% yield of press mix. Evaluation of the spray-dried mix is by how well it fills the die cavity and how well the granules coalesce during pressing. This is determined by the density of the fired part and by microscopic examination.

Stir Drying

This is done in a mixer that has a heater. As the material loses water, it is continuously mixed to keep the constituent's distribution uniform. This is a very common technique in the chemical industry. Steam jacketing is a common practice. Commercial equipment is oversized for the lab, but one can make a lab-sized, stir drier. For instance, one can fabricate a heating mantle for a Hobart mixer. This works well except that one has to periodically scrape down material that sticks to the sides of the bowl. Vendors of heated driers will run tests on their equipment. The slip can form a dilatant mass after it has lost some water. (This can overload the stirring motor and bend the stirring mechanism into a pretzel.)

One thing to remember is that evaporation of the water is endo-thermic and the mix will be cooled during drying. When open to the air, condensation will occur in the mix unless it is heated. The worst case is a mix with a high vapor pressure solvent such as methylene chloride. The mixer will become so cold that a thick layer of frost will form on the bowl. One can imagine the amount of water that is condensing in the mix.

One needs to scrape down the bowl and blade regularly during drying. This hardened material will form agglomerates in the mix, but will have the same composition as the powder.

The blade (as in the Hobart) will not conform closely to the sides of the bowl creating a dead space. Rubber scrapers can be fastened to the sides of the blade and act as squeegees. It will still be necessary to scrape down the parts.

Solvent Drying

As the mix loses the liquid, menisci draw the particles together by surface tension, agglomerating the mix. Interfacial tension with liquids other than water is lower and this effect is reduced. In solvent drying, the starting liquid is probably water that has to be replaced with the solvent. The first step is to recover the solids. Methods to do this include: sedimentation, filtering, centrifuging, or just pouring the slip out onto a flat slab of plaster, which is an old potter's trick. Then, the material is re-blunged with a solvent such as acetone or methanol. As water is miscible with these solvents, it is diluted and is decanted along with the solvent. A repetition of this process will further reduce the water content. The powder can now be dried. Alternatively, it can be reblunged with a nonpolar solvent with even a lower interfacial tension, such as a low molecular weight hydrocarbon often containing a wetting agent.

Again, these solvents are flammable and some are toxic, so safety should be considered. An oven can be converted to be explosion proof by removing the thermostat relay to a remote position and plugging the oven into a variable transformer to reduce the element temperature to as low as is practical. One should still have a good air flow to keep the solvent vapor concentration low. Be sure to check the voltage each time to make sure that somebody has not turned up the dial. When there is a substantial amount of solvent drying, the oven should have a hood and an explosion panel directing the explosion away from people. Alternatively, one can let the solvent air dry. When the amount is small, this can be done on a table top in a ventilated room. As before, there is the problem of water condensation.

Rotary Drying

In this process, the wet material is fed into a pitched tube that is slowly rotating. Heaters raise the temperature to where the water is evaporated, at which point the dry mix falls out of the other end. Fine-grained materials stick in the tube and coat its surface. In some production instances, chains are hung in the tube to break up the cake. Agglomeration occurs as one would expect. One can place a high-intensity, dry mill at the exit end of the tube to mechanically break up the larger agglomerates. One does not apply this process to powders containing binders or other additives. Ordinarily but not always true, rotary driers are used in the plant rather than in the lab. A major problem is adherence of the cake to the walls of the tube.

Freeze Drying

This process has the potential of being very useful as the water is removed by sublimation and this does not result in agglomerates, at least from the drying process.

Steps in the process include pouring the slip into shallow trays, placing them in a freezer, and subliming the water. The temperature of the tray has to be higher than that of the cooling coil to transfer the water to the coil. Trays are heated but have to be below freezing or the slip will melt. To control the temperature differential, the slip has to be in a thin layer and all of the temperatures have to be below the freezing point of water.

Other freeze drying techniques involve spraying the slip into liquid nitrogen, alcohol cooling with dry ice, or acetone cooling with dry ice. The spherical particles are then recovered, kept cold, and freeze dried. One will need to screen the mix as spray nozzles produce a wide range of sizes.

The industrial process for commercial freeze drying is done at a huge facility in the Midwest. The problem is the restriction of the thin layer. This cuts into productivity, unless the market is large and the facility is automated. Instant coffee is sometimes freeze dried.

Freeze drying is a valid laboratory procedure for drying a slip and producing a powder. Laboratory-sized, freeze driers are available. One should anticipate a scale up to production. One should also inform manufacturing if the lab is successful as the plant will invest in freeze-drying equipment.

Check List, Drying

- Pan drying is the most commonly used technique, but it leads to segregation and agglomeration.
- Spray drying, at present, is the preferred technique for fine particles despite all of its faults.
- Spherical particles are a notable advantage in mold fill.
- Segregation is a serious problem in drying. Mix while drying.
- Agglomeration is a serious problem in drying. Use volatile solvents.
- Coarse-grained mixes can be pan dried as segregation is not as much of a problem, provided that the mix is only damp.
- Freeze drying is a good lab method to prevent drying agglomerates.
- Controlling the uniformity of the powder is not easy, and much of this results from the drying process.

3.0 GRANULATION

For die cavity filling, the press mix should be uniform and free flowing. When it is a fine stiff cake or formless mass coming out of the mixer, granulate the material. One type of granulator is a U-shaped screen with an arm that oscillates back and forth to press the mix through the screen. The mix has to be damp and with a binder. Both the amount and viscosity of the binder solution are critical and determined for each material. The binder solution has to stick particles together to form the granule, so it seems that a viscous liquid is preferable. The same process can, of course, be done with a lab screen by pressing it manually through with a plastic or rubber scraper. While it depends on the particle size, the screen is in the range of 40 to about 60 mesh. As the mix is pressed through the screen, it forms granules consistent with the mesh size. This can now be dried, usually aired or pan dried.

Some rotating mixers are capable of granulating the mix. Granulation occurs at a slower speed than mixing with the speed being a critical factor. Problems are with forming lumps or a loose powder. The equipment manufacturer can probably help with finding the right conditions and binder types.

Usually, this is more appropriate for pelletizing than granulation. Rotary drums can be used where the dry mix is sprayed with the binder solution while it is tumbling. Just at the right point, the mix will start to ball up, sometimes making pellets and less often making granules. After drying, the mix is screened to narrow the particle size distribution.

Check List, Granulation

- Use of a granulator
- Screening
- Binder type and amount critical

4.0 DIE PRESSING

Die pressing is a very widely used process for forming ceramics. It is suitable for both fine and coarse grained ceramics. This section is divided into eight parts: drying, granulation, screening, die manufacture, basic shapes, cavity fill, pressing procedures, and problems with pressing.

Die Manufacture

Die tolerance, which is the gap between the punch and the die body, is tighter for fine press mixes than those for coarse mixes, which seems reasonable. Usually, 0.001" is common for fine mixes, 0.002" is common for coarser mixes, but sometimes this can be greater. For lab applications, dies are usually relatively small and can be manually moved. Usually, lab dies are made from hardened steel, while production dies sometimes are faced with cemented carbide. Durability is not ordinarily an issue in lab tooling as the number of parts is limited. Both metallurgy and machining will be discussed.

Metallurgy

One can use a variety of tool steels for dies. The best source of information is the die machine shop. There are two requirements for the metal: hardness and elongation to fracture. Hardness determines the die wear resistance, and is usually between 55 and 60 Rockwell C. Since the die wear mechanism is abrasion, harder metal wears at a lower rate. Hardness is limited by brittleness. There is a relationship between hardness and brittleness: harder is more brittle. Some elongation to fracture is needed to give the metal some forgiveness. At Rockwell C 65, the metal is said to be *glass hard*. With no give, the die may break, and this can be hazardous and costly. When one needs a very hard die, an outer shell of soft steel can be shrunk fit on forming a safety ring. The press should have a safety shield, regardless. The elongation to fracture should be around 10 to 15%. ASM International Metals Handbook gives this information and the heat treatment necessary to attain the desired properties. [1]

Not everyone realizes that the modulus of elasticity does not change appreciably as the steel is hardened. While it becomes harder, it does not become stiffer. Multi-component dies are often fastened with bolts. Always use high strength bolts or cap screws. The tensile strength of these is between 150,000 and 170,000 psi. The cap screws are marked on top as to the strength class; six markings are the high strength ones. Cap screws with a recessed head result in a flush surface and are easier to handle. Stainless steel galls are more difficult to work with. Some stainless steels can be hardened, but they are not commonly used because of the tendency to gall. Anti-galling compounds are available. These contain copper and aluminum flakes in grease. This is not the sort of thing one gets into the press mix. If it is necessary to use stainless, use the compound or the die will be ruined. The compound can be found at an auto parts store. Now, it is not necessary to know all about metallurgy to ask the machine shop to make a die. If one asks the right questions at the machine shop, they will be a little more careful with making the die. Shop people like to talk to knowledgeable clients as it enhances their own status and gives them an opportunity to contribute. These craftsmen/women have impressive skills.

Machining Dies

The shop will use milling machines, drill presses, lathes, or other metal cutting tools to rough cut the die shape while the steel is still soft. They will cut a little fat to finish the die to tolerances after it has been hardened. Machine tools usually produce surfaces that are planer, right circular cylinders, or conical. Well-equipped machine shops can produce a variety of other shapes especially if they have computer-controlled, contour facilities. After hardening, the die parts are ground to the final tolerance and finish. We should discuss a few topics on machining.

Grinding Square Corners. A grinding wheel cannot reach into a square corner, as shown in Figure 6.3. Part A of the sketch shows where the edge of the wheel jams up on the corner. Part B shows how this can be relieved by a groove at the corner. These surfaces often are alignment surfaces and have to be flat and square. Another case is machining an inside cone, as shown in Figure 6.4.

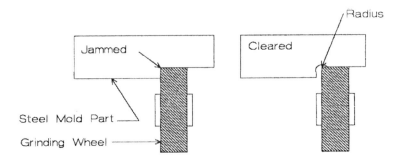

Figure 6.3: Grinding to a Corner. The grinding wheel will jam in the corner unless relief is provided.

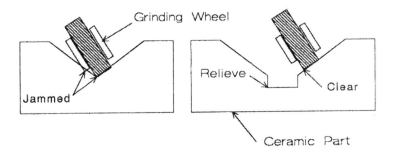

Figure 6.4: Grinding a Conical Surface. Relief must be provided at the apex to prevent jamming.

This is a similar problem. The grinding wheel cannot reach into the apex and has to have a run-out space in order to grind the conical surface properly.

Bolt Positioning. We have placed three bolts at 120 degrees around a bolt circle. There is a better way, where the bolts are placed such that the die body can only fit one way, always the same. One way is shown in Figure 6.5.

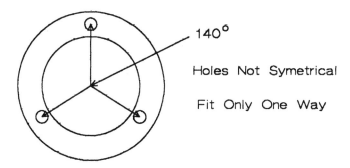

Figure 6.5: Offset Bolt Positioning. Offset fasteners provide only one assembly configuration.

This is only one example of a foolproof design. There are many variations of the same principle. Design the assembly so that it can only go together one way. Electronics designers do similarly when designing connectors. A similar thing can be done with plates. Consider the front and back plates of a die as shown in Figure 6.6. The bolts are placed a little off center so that the plates can be assembled only one way. It is also helpful to mark the corners for a visual reference.

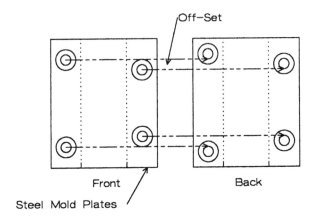

Figure 6.6: Offset Plate Bolting. Offset bolts are designed so that the die is assembled only one way.

Die Parts Positioning. This is another example of foolproof design. Rectangular dies are fastened with bolts that are not precision positioners. Pins can be added and placed so that they fit only one way and with precision. Figure 6.7 shows such an example. Looking back at Figure 6.3, the properly ground flats will assure that the mold is assembled rectilinearly, and the pins assure that it is aligned laterally.

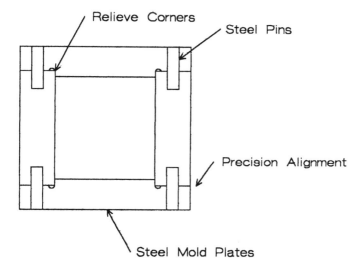

Figure 6.7: Positioning with Pins. Pins are a precise way to provide alignment.

Fastening to Shafts. Parts can be threaded onto shafts. If properly done, the thread will not unscrew during use. There are both right-handed and left-handed threads. Use a thread where the torque on the shaft will tighten the thread. One can also use jam nuts to hold the part in place. Devices such as cotter pins or castellated nuts can prevent the thread from unscrewing. When there is not much space, one can use tapered pins as the fastener through a close fitting sleeve. Figure 6.8 shows such an example. Tapered pins result in a flush surface that can be a safety advantage, as well as conserving space. They, when selected at the appropriate size, can also act as shear pins that limit the stresses on the shaft.

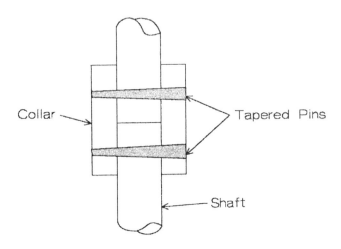

Figure 6.8: Fastening a Shaft with Tapered Pins. Fastening is secure and compact with pins.

Strength. Holes and notches are stress risers and will weaken the die. Whenever one can, it is better to put a radius on inside corners and position the holes where they are not in the high stress areas. A mechanical engineer can calculate the stresses in the die design under load. For lab dies, this may not be necessary as it is easy to over design and to avoid the problem.

Machine Shops. These can be internal or external. Internal shops often have their own priorities, especially if they are also doing work for production. Additionally, if the project is not perceived to have a high priority, the project will be placed at the bottom of the list. These priorities are capricious much of the time, and it can be a battle to get the work done. However, there are several alternatives. When it is critical to assign your job to an outside machine shop, assign a machinist specifically to your project. The outside shop is far better than the other choices as the outside machine shop does not get paid for the job until they furnish the part correctly. Money is a very powerful incentive.

The other advantage is that there are choices for selecting particular skills. Not every shop has all of the same skills and it is worth while to shop around.

Finish. Hardened steel dies are machined oversized, heat treated, and ground to finished dimensions. Faces on die punches should be polished to a reflective finish, especially for fine-grained, press mixes. It is also better to have the inside surfaces polished. There might be some difficulty in finding a shop to do this work locally. There are advertisements in the Bulletin of the American Ceramic Society for this work.[2] A superfinish is not a reflective polish, but instead is a lapped finish that is not the best choice. One can coat die surfaces with many materials by applying many processes. Electroplating with chromium is common, but one cannot plate the inside unless the die is dismantled. Plating the outside does slow rusting a little bit. Punch faces can be plated with hard chrome, a reflective finish, provided the steel is previously polished. There are a variety of CVD coatings to coat the die face. CVD coatings are applied at elevated temperatures. CVD materials have a high modulus deposited on a lower modulus substrate.

Basic Shapes

Laboratory dies usually have simple shapes. Some of these shapes and the die setup in the press are discussed here.

Right Circular Cylinders

The die for making this part has four components, as shown in Figure 6.9. With the top punch out, the press mix is poured into the die cavity and leveled. There are two ways to meter out the amount of press mix: by volume or by weight. When metering by volume, the mix fills the cavity and is screed off the die top. Then, the lower punch and mix are lowered by removing an additional shim (not shown) and the top punch is then inserted.

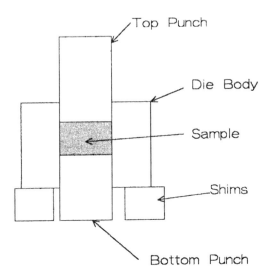

Figure 6.9: Cylindrical Die Assembly. The assembly includes a die body, two punches, and shims.

A fill by weight is done by pouring the mix into the die body and leveling it. Leveling is accomplished by vibration, tapping, or with a leveling tool, which is preferred. A leveling tool is rotated in the die to spread the mix out evenly with a blade. The blade is gradually raised to where no more mix is plowed. This is shown in Figure 6.10.

The blade edge should be a knife edge so that the mix is not compacted during leveling. When leveling in a rectangular die, always start from the end and work toward the center to avoid packing the mix at the ends. This is useful whenever the part shape is critical. Packing will lower the firing shrinkage locally and distort the part.

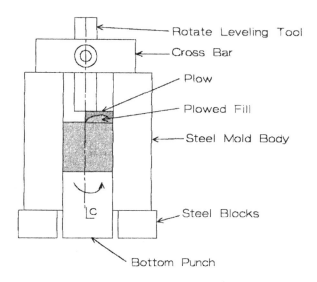

Figure 6.10: Mix Levelling Tool. A weighed amount of powder is levelled by rotating the plow and slowly raising it to where all of the mix is level.

Referring to Figure 6.9, after the mix is leveled, the top punch is inserted and the die is pressed to part of the final load. Friction between the pressing and the die wall is enough to support the die as the pressure is released. This is called *bumping the die* and allows entrapped air to escape, and the shims to be removed. Do not put fingers under the die body at this stage as the friction may not be enough to support the die. Since both the top and bottom punches are free to move, the pressing is now double-end pressed to the full pressure. Double-end pressing is far superior to single-end pressing. Pressure gradients in the pressing are much less with pressure applied to both ends, and the part will sinter with less distortion. These gradients were described by Kingery[3] and later by Thompson[4] in an American Ceramic Society Bulletin. Density gradients are especially severe when the part is long with respect to its diameter. Pressure is held for a few seconds as the punches will continue to compact the part. A dial micrometer on the press platen will display this movement and is a good guide about when the pressure can be released. The part can now be stripped from the die. The die pressing process is shown in Figure 6.11.

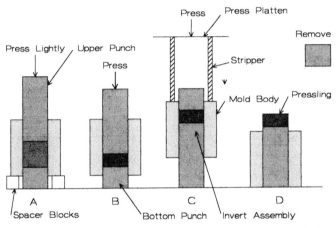

Figure 6.11: Stripping the Die. There are four steps: A. fill the die and insert the top punch press lightly, B. remove the shims and press the part, C. invert the die and place the stripper on top, and D. press down the strip part out of the die.

Part A is the die with the fill ready for pressing. Spacer blocks hold up the die body. B shows the die after lightly pressing and with the spacers removed. Full double-ended pressure is now applied. C shows the die turned over and the stripper placed on the assembly. Strippers can be blocks or a square length of pipe with one side cut out. It must be slightly longer than the bottom punch plus the sample or the sample will be crushed. D shows the part after stripping. To lift the part off cleanly, the top punch has to be a little longer than the die body. During this operation, the die body may suddenly drop. It is not a good idea to place one's fingers under the die body.

There are a number of problems when die pressing. These will be discussed after other die configurations are considered, as the problems are common to all. Refrain from cutting metal until these problems are considered.

Flat Plates

Dies for pressing flat plates are assembled from hardened steel blocks bolted together. Figure 6.12 shows two such illustrations. The design on the right has the advantage that there are only two shapes to the machine. If the cavity is square, then there is only a single shape for all the pieces, saving money in machining costs. Bolt positioning is not optimized in this design for two reasons. First, the bolts are in shear from pressure on adjacent plates. Secondly, the adjacent plates are not pulled up tight on the alignment surfaces. Figure 6.13 depicts a better design. There are two alignment surfaces at each corner A and B. The bolts pull the pieces together against both surfaces that have been precision ground. Each bolt holds the plates with tensile forces. The bolts are at different levels and should be staggered so the die goes together only one way. Two or three tiers of bolts are sufficient for most lab dies depending on the length of the piece being pressed.

Punches for rectangular dies can be a little heavy, and sometimes can be an assembly of a plate that fits the cavity bolted to an extension with a smaller cross section. It is a good idea to hone the sharp corners off the die parts to avoid cuts to one's hands. This can be done quickly with silicon carbide abrasive cloth, available at the local hardware store.

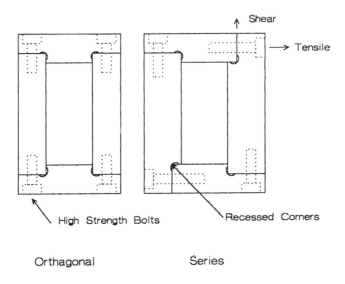

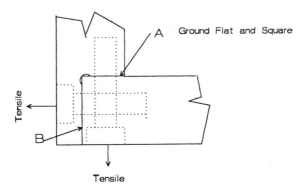

Figure 6.12: Rectangular Die Assembly. There are two ways to design the configuration: orthogonal and series.

Figure 6.13: Die Corner Detail. Each corner has bolts that are in tension in both directions for stability.

Compound Heights

Often, the part will not have a uniform cross section in the vertical dimension. An example is shown in Figure 6.14.

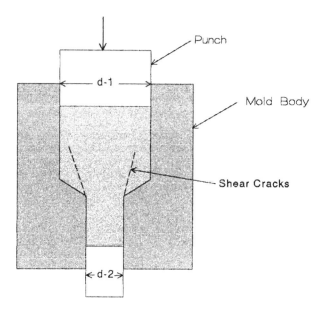

Figure 6.14: Shear Zones in Compound Heights. The center will compact to a lower density than the edges producing cracks and a low density center.

While this example is hypothetical, others just as problematic exist in the real world. There are two problems: length in the center is longer than the length around the sides, and the piece cannot be pressed evenly with more pressure on the shorter side sections. Shear occurs and the piece will likely crack at the location shown. Differences in packing density lead to differences in firing shrinkage. For a shape this difficult, it might be necessary to mud press, and there is no guarantee that this will work. It may work on a coarse- grained mix but not on a fine-grained mix. Mud pressing uses a mix with enough water and rheology modifiers to make it plastic. In a case like this, the part had to be set on a plaster form to draw off enough water to where it would not slump after stripping from the mold. When the difference between the central length and edge length is much less, there is a chance of pressing the part.

Rings

Rings can be die pressed with tooling as shown in Figure 6.15.

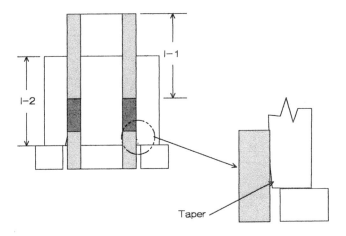

Figure 6.15: Pressing Rings. Rings are fragile. A taper on the corner of the die gradually relieves stresses during stripping.

The central part is floating, or can be held shimmed in place. The problem with pressing rings is that they can be fragile. Spring-back will cause the OD to expand as the ring is stripped from the mold. Note that Figure 6.15 has a relief taper on the bottom of the die. In case there is breakage, the angle of the taper and length should be increased to where the spring-back is compensated. Spring-back on the ID releases the part from the core so that it slips out easily. Fine-grained pressings are more susceptible to fracture than coarse-grained pressings because they have more spring-back. High pressure increases spring-back, and the part becomes even more easily damaged. Backing off on the pressure can help reduce breakage.

Check List, Pressing dies

- Hardened tool steel
- Alignment surfaces
- Die foolproof assembly
- Die strength
- Choice of machine shops
- Polish
- Die design, cylinders, plates, compound heights,
- Rings

Hydraulic presses

In the forgoing discussion, it was assumed that the hydraulic press was simple with only one ram. Of course, presses are available with lots of additional useful functions. A double-ended press, with equal top and bottom rams, and a stripping ram allows the use of fixed dies fastened to the stripping ram. Then, it is not necessary to consider stripping blocks and handling the die. If affordable, one can get fancy control systems where the pressing conditions are automatic.

A 50-ton press is about the right capacity for lab projects. There are some necessary controls including: force on the rams (hydraulic pressure), volumetric flow rate of the hydraulic fluid into the cylinder, dwell time, and stripping rate. Platen positioning is very useful to avoid frequently running the press up and down. Gages dampened with glycerin prevent the needle from jumping around to where it becomes impossible to read accurately. A press this size is preferably a four poster. C-frame presses are tolerable if they are of lower capacity and the additional access is needed from three sides.

Smaller bench presses are useful and available. These are essentially a car jack mounted on a base plate, two posts, and an adjustable top platen. A pressure gage and a hand pump comprise the hydraulic system. These can also have a hydraulic system, and should definitely have a safety shield. Safety shields should protect the operator and others in the lab.

Press platens are not always parallel, especially the type of press just described, as the top platen is frequently moved. One can use a simple alignment fixture to compensate for this misalignment as in Figure 6.16.

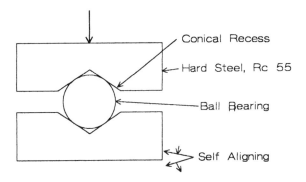

Figure 6.16: Die Alignment Fixture. A ball and cone fixture aligns the pressing assembly.

The fixture is placed on the top punch. It is a good idea to make sure that the top platen is well aligned so that the fixture does not squirt out under pressure, although this is not a problem. Higher pressures require larger fixtures, but for most lab work a ball bearing between one half inch and one inch D is about right.

Problems With Die Pressing

Many common things can go wrong. They are:

Lamination

Laminations are cracks that run parallel to the press platens close to the center of the piece. These are usually caused by entrapped air. There are a variety of remedies that usually work. Sometimes, the lamination does not run all the way to the surface, and the piece will have to be broken open to view it. Do not do this on every piece.

Bump the press. After pressing to about 80% of the punch travel, release the pressure and allow the compressed air to escape. For this to work, the pressing has to be permeable, which is a consequence of the green density.

Segmented Punch. When pressing flat plates, the air has to travel a long distance to escape and laminations are common. When the punch face consists of several parts, air can escape along the cracks, as shown in Figure 6.17.

Evacuate the cavity. By removing the air with a vacuum pump, the air entrapment is avoided. The pump can be connected with a side port as shown in Figure 6.18.

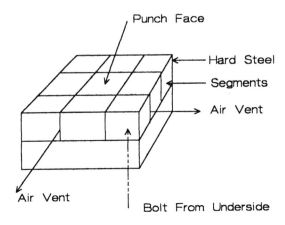

Figure 6.17: Segmented Punch Face. Air can vent retarding laminations.

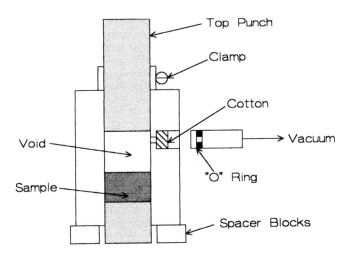

Figure 6.18: Evacuated Die. Removal of air prevents laminations.

The side port has to include a filter or the mix may end up in the vacuum pump. A wad of cotton works fine when it is stuffed into the side port. When starting to evacuate, the top punch is held above the port. As the vacuum is pulled, the top punch will immediately drop down onto the mix. To fix this, simply tighten a worm gear hose clamp on the top punch against the top of the die body. After removing the air, the clamp is loosened and pressing can begin. For a square die, other types of clamps are effective. A high vacuum is not necessary. All that is required is to remove most of the air, preventing internal air compression sufficient to cause lamination.

Edge relief. This has been mentioned earlier, where a chamfer is ground on the bottom of the die cavity so that spring-back recovery is gradual during stripping. This type of lamination is caused not by air pressure but by mechanical shear stresses.

Mix Consistency. For purposes of illustration, consider a spray dried press mix that is being compacted. When the mix is plastic, it will squash together and seal off pockets of air. When the mix is brittle, the granules will fracture and the pressing will remain permeable allowing the air to escape. Depending on the binder system, the plasticity can be adjusted by changing the formulation, often the water content. Some ceramists use bone dry press mixes. Figure 6.19 is of a sintered body where the relic granular structure of the spray dried spheres is evident.

End Capping

This is a crack running in from the top edge of the pressing at an angle to the center, as shown in Figure 6.20.

Figure 6-19: Relict Spray Dried Structure. The spray dried spheres did not coalesce during pressing. Scale bar 100 μm.

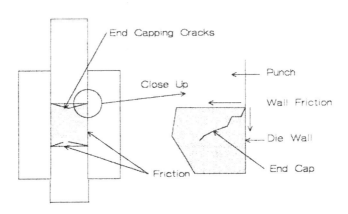

Figure 6.20: End Capping. Cracks on the end of the piece are produced during pressing.

There is an analysis by Thompson on end capping.[5] It is caused by frictional forces across the punch face and along the die walls. It is also affected by the press mix consistency, with fewer plastic mixes tending to end cap. Decreasing friction between the sample, die, and punch faces can be accomplished by polishing. In the lab, the punch face can be lubricated for each pressing. Try different "lubricants" (the quotes suggest such substances may not necessarily be thought of as lubricants). Stearates can work and are worth a try. In the plant, it would not be practical to lubricate the die for each pressing.

Sticking

Press mixes often stick to the punch faces, causing a reject. Then, before proceeding, the punches have to be cleaned; maybe twice. Solutions to the problem in the lab are not always available in the plant because they can be time-consuming.

Polishing. Polishing the punch faces can help both in the lab and in the plant. A mirror-like, reflective finish is required. The surface will have a minimum area for adhesion, reducing the tendency to stick.

Paper Divider. This is an infallible lab procedure that eliminates sticking. For a round die, a disc of paper is placed on both punch faces to provide a release surface. A paper punch can be made so that the disc fits just right. Make the disc a little undersize so it will go in easily. Figure 6.21 depicts a paper punch design.

Always use the punch on a plastic or wood block. A sharp blow with a mallet will always cut out a perfect disc. In order to preserve the disc diameter, sharpen the cutter by grinding the beveled face.

Release Agents. Some release agents expected to work, fail instead. Silicone release agents do not work well for release. Stearates sometimes work. The best agent for oxides is polyethylene glycol 400 that one applies on the punch faces for each pressing; however, this would be prohibitive in the plant. A more general approach is to incorporate a release agent in the formulation. Try stearates and try changing the binder or plasticizer.

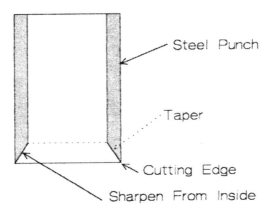

Figure 6.21: Paper Punch. This punch is useful to cut out precise discs for use in part release.

Hour Glassing

Due to the density gradients in the pressing, the center will have a higher firing shrinkage than the ends. The fired shape resembles an hourglass, or a figure with a waist. When the pressing is single-end pressed, the shape is exaggerated. Hour glassing is a natural consequence of the process. It can be reduced by the following: using a more plastic mix, pressing to a higher pressure, or following pressing with isopressing and green machining. There is additional discussion about this in the chapter on firing.

Warping

Any sporadic difference in green density will result in warping or distortion of the geometry. Warping can also occur when there is a temperature gradient across the piece during firing, where the hot part sinters more than the cold part. However, the main cause of warping is green density gradients due to an uneven fill. If the kiln has a cold hearth, change it. Raise the hearth so that heat can get under the part as it is firing. The setting technique of the part on the hearth can also cause distortions. This will be discussed in the chapter on firing.

Cracking

Cracks are usually nucleated in pressing. This is usually due to the following factors: shear from uneven cross sections, corners in the die, or differences in thickness. Fine-grained ceramics are also susceptible to thermal shock in firing. The crack is usually formed before the soak. When the crack edges are rounded, the crack is present before sintering. Another factor is that the body is very weak after the binder is burned out. With a very low strength, the part can be easily damaged. Some cracks do not appear on the surface and can only be seen after the part is sectioned. Such cracks are the worst kind as they will become noticeable only after evaluation. The following changes will help eliminate cracks: the tooling geometry, pressing rate, binder/plasticizer, binder strength, pressure reduction, and slowing the temperature rise when firing through the critical region between binder burnout and the start of sintering. In desperation, change the forming method.

Check List, Pressing Problems

- Laminations: bump the press, segmented punch face, vacuum evacuation, stripping edge relief, less plastic mixes
- End Capping: polish the die, less plastic mixes, die lubricants
- Sticking: paper divider, polish the faces, release agents
- Hour glassing: double end press, more plastic mix, higher pressure
- Warping: hot hearth, die design, die fill
- Cracking: die design, even fill, binder/plasticizer, lower pressing rate, stronger binder, slower binder burn change forming method

5.0 OTHER PRESSING TECHNIQUES

Hot pressing and hipping are covered in the chapter on firing. The discussion here is on isopressing.

Isopressing

There are two kinds of isopressing: wet bag and dry bag. In both cases, the press mix is placed in an elastomer bag and compacted with hydraulic pressure on the exterior of the bag. Wet bag isopressing is generally used in the lab as dry bag has dedicated tooling for a particular size and shape.

Wet Bag Isopressing

Tooling consists of a rubber bag into which the press mix or preform is placed. The bag is sealed and placed in the isopress. There is a pump that increases the pressure in the press, on the bag, and on the mix; this causes it to compact. Figure 6.22 depicts a photograph of such an isopress.

Features of the isopress are a top closure, a pressure vessel, a diaphragm pump, a fluid storage tank, and a pneumatic lift. The fluid is water with a rust inhibitor. After a bag ruptures by accident, the fluid may become contaminated with ceramic particles that are abrasive and require draining and flushing of the system. Various capacity air pumps are available, with the choice being between speed and cost. For lab work, speed is not too critical.

One makes a rack to hold the samples; this rack is lowered into the cavity. With a loop on top of the rack, it can be snagged with a hook and lifted out of the press. Since shop air may not be clean and dry, it is advisable to filter the air and to add a trap.

After one acquires an isopress, one should choose a large enough ID to adapt to future needs. For general purposes, an ID of between 4" and 6" is about right. Keep in mind that as the volume increases, the pump capacity also increases by the square of the ID radius. There may be a need for a larger pump.

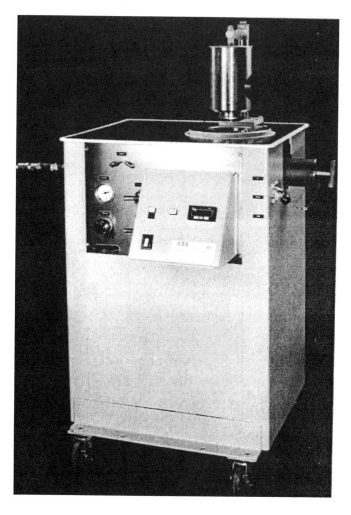

Figure 6.22: Cold Isopress. Powders or preforms can be uniformly compacted in an isopress. (Courtesy of ABB Autoclave)

One of the principal advantages of isopressing is that the pressing is uniformly compacted, which results in more uniform sintering. Figure 6.23 shows such a typical tooling setup.

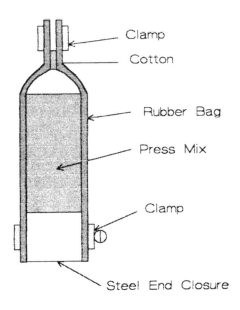

Figure 6.23: Isopress Tooling Cylinder. A rubber bag and end closures contain the sample during isopressing.

Basic tooling consists of the following parts: a rubber bag, a base, a clamp, and a wad of cotton. To seal the interior from the liquid, fasten the bag and base with a twisted wire or electrical tape. In the vibrating mode, fill the bag to the base of the neck with press mix. The cotton wad is then pushed down into the neck to prevent aspiration of the mix into the vacuum pump. A vacuum is drawn through the neck and then the mix is sealed with a threaded hose clamp. One can then insert it into the isopress. There are two types of isopress closures: breach block and top plug with a lateral pin. Either closure will do. The plug or breach block is lifted with an air cylinder.

Pressing a closed end tube is a source of unexpected difficulty. The tooling is shown in Figure 6.24.

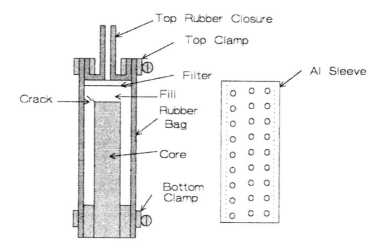

Figure 6.24:   Isopress Tooling, Tube. Pressing around a central rod produces a tube.

The core rod is shown with a square end; however, it is rounded. A square end will always crack at the corner due to shear stresses. The top closure consists of a rubber sleeve with a connecting tube that is folded back on itself. A steel ring provides a structure to clamp against, and another clamp is used to seal the bag when isopressing.

The setup is the same except that there is a central rod to form the ID. Using the same procedure, the tube is pressed at 60,000 psi. When pressing to this high pressure, the rod can bend during the run. Substituting a hardened steel rod, the rod can still bend. The reason for this is obscure. One can circumvent this problem by doing the following: using a spider to center the rod during fill, encasing the bag with a perforated aluminum cylinder to hold it in shape, and lowering the pressure to 20,000 psi. The setup is sketched in Figure 6.25. The aluminum cylinder is placed around the rubber bag during fill and isopressing.

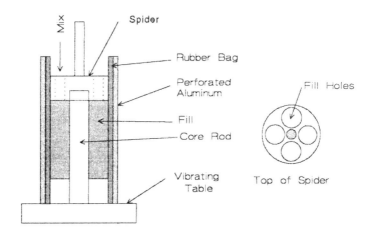

Figure 6.25: Spider for Centering. The spider centers the core rod for uniform fill. The aluminum cylinder holds the rubber bag in shape.

When there is close control of the firing shrinkage and the mix is pressed around the rod, the center hole in the part is very precise.. The outside diameter of the part is irregular with this tooling and will need to be green machined. Shapes with other cross sections can be made in a similar way. Some rubber companies besides provide sound advice and specialize in making whatever isopress tooling is needed.

Condoms. Unless convenient, it is not necessary to have special tooling to isopress. A pressing can be placed in a condom and simply tied off. It is necessary to remember that condoms are expensive, so it is more economical to buy in bulk directly from the manufacturer. Also, condoms come with a variety of coatings, including lubrication, silica dust, and talc, but they should be ordered cleaned and not rolled. There can be a problem with condoms leaking as well. Often, leaking is caused by air pockets that pop. Inspect the condoms to see if they have holes. Vacuum de-airing helps, but it may become necessary to double or triple bag or to change suppliers.

Heat Sealed Plastic Film. Another trick to bag a pressing is to encase it in a bag of heat sealed plastic film. The bag can be sealed around three edges. Slip the pressing into the bag and seal the third side, except for one corner. A vacuum nozzle is pushed into the corner and the bag is evacuated. With the nozzle still in place, the corner is sealed shut while still pulling a vacuum. It can now be isopressed.

Latex Isopressing.[6] One can dip a pressing in a latex emulsion forming a rubber bag around the part when dried. However, when the part displaces air by capillary intrusion of the latex, bubbles can form in the film. To prevent this, the part can be sealed first with a polymer in solution. This is similar to a sanding sealer used on wood. Then, the part is dipped in the latex and dried. This will work at room temperature or at 60 °C. It is a good idea to double dip. One can also use rubber cement (office type), other elastomers, or deformable materials that will not crack. There are some advantages to this process: any preformed shape can be isopressed and parts can be isopressed in bulk. When bulk isopressing, the preformed parts are placed in a mesh bag and then dipped into the latex. The bag is like the type used to contain potatoes or onions in the supermarket. Volumes of space can become isolated during drying of the emulsion and will crush in the press. To avoid this, dilute the latex with one part of water to two parts of latex. Because the latex has a pH of 10.3; it is a good idea to adjust the water to this pH before it is added. Dilution lowers the viscosity of the latex, allowing it to drain after dipping. There are latex emulsions for making surgical gloves and latex emulsions for industrial uses. The latter are much cheaper and more satisfactory.

Dry Bag Isopressing

The difference between wet and dry pressing is that in dry bag pressing the rubber bag is an integral part of the apparatus as shown in Figure 6.26.

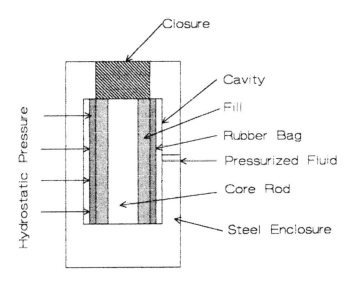

Figure 6-26: Dry Bag Isopress. The rubber liner is fastened to the isopress structure. A central core rod will produce a tube.

Dry bag isopressing is especially useful for making shapes such as tubes, spark plugs, and crucibles. It is a very useful production tool, but since the tooling is so dedicated to specific sizes and shapes, it is not commonly seen in the lab. Procedures are to fill the cavity with the press mix, close the top, and apply pressure to the rubber sleeve. The pressure is released and the part ejected, often with an air cylinder.

Check List, Isopressing

- Elastomer bags
- Perforated Al tube support
- Vacuum de-air
- Do not overpress
- Spider to center central rod
- Double bag condoms
- Order clean and flat condoms
- Heat seal plastic film bag
- Latex isopress complex shapes and bulk materials

6.O SLIP CASTING PROCEDURES

It was observed, on a microscopic scale, why slip casting can sometimes produce a higher green density than die pressing, isopressing, or injection molding. A dilute water suspension of alumina was being observed on a slide. Particles were migrating toward the edge of the slide where a crust was forming due to drying. This crust was analogous to a permeable plaster wall adsorbing the water from the suspension. As the particles approached the wall, they streamed in channels and oriented with the long dimension parallel to the channel direction because of fluid shear across the channel. As the channel narrowed, the particles wedged into the wall with the orientation geometry, resulting in dense packing. As one channel became obstructed, others would open and the process would repeat. High green density in slip casting is the consequence of fluid mechanics, which orients the particles and jams them into the casting surface. Under ideal slip casting conditions, green densities up to 60% can be achieved on sub-micrometer slips.

Leading up to this section on casting procedures, several topics were discussed earlier. These include slip preparation, binders, slip storage, and the effects on viscosity of the surface chemistry.

Plaster

Plaster of Paris is a mineral product made from calcined gypsum in which one half of a molecule of water is left in the structure. If lacking all the water, the material becomes anhydrous and is useless for casting. Number One pottery plaster is generally used for slip casting. There are other grades that are useful for other things, but use Number One for slip casting.

Plaster of Paris will absorb water when exposed. To slow the hydrating process, store plaster in a plastic bag, place the bag in a garbage can, and cover the can with a lid. Plaster will keep for a few months if kept from hydrating. Figure 6.27 shows the effect of consistency on strength, absorption, and dry specific gravity.

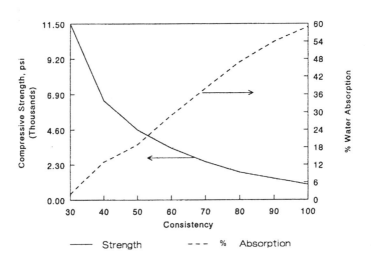

Figure 6.27: Plaster Consistency. As more water is added, the plaster strength decreases and absorption increases.

Consistency refers to parts of water per 100 parts of plaster by weight. For slip casting, a consistency of 60-70 is in the right range. Almost all plaster shops use tap water at room temperature or, control the temperature so that it is always the same. Water fresh out of the tap will not be at room temperature, so a holding tank is a good idea. Deionized water from a holding tank for lab use is good practice because it is consistent. If one uses tap water, filter it with a fine filter. Distilled water is available at the supermarket and is an alternative. Read the label and make sure it is distilled, however. Spring water may not necessarily be distilled water. Dried plaster is porous so that it can absorb water. Figure 6.28 is an SEM photo of a fracture surface on plaster.

Figure 6.28: Plaster Microstructure. Plaster forms clusters of needles with porosity in between. Scale bar 10 μm.

The structure is comprised of interlocking needles of the plaster. Interstices between the needles are the channels through which the water flows during casting.

Process For Making a Plaster Slurry[8]

Formula. To obtain a consistency of 67, mix two parts of water to three parts of plaster by weight. This results in 0.065 final pounds of plaster per cubic inch. The formula establishes how much to mix provided the volume of the mold is known.

To avoid clumps, sift the plaster powder through the fingers into the water. Allow the plaster to slake for three and a half minutes. Next, stir with a mixer, at a moderate speed, for one minute. The plaster will start to hydrate. To prevent settling of the plaster particles and to remove large air bubbles, stir the slurry slowly by hand in a lifting motion. The viscosity will increase in a few minutes to where a path made by dragging your finger across the surface will persist. It is now thickened and ready to pour into the mold. If one pours it too soon, the particles will settle. When poured too late, the plaster will not completely fill the mold.

Revised Plaster Process

The traditional slip process has worked for decades, and many plaster shops still use it or a similar version. Some experience with slips suggests a new process where the plaster is mixed at a higher intensity to deagglomerate the plaster particles. This results in a rapid set. High intensity mixing releases the particles from the agglomerates so that they can participate in the setting process. Only the free particles can participate in the set. Plaster concealed within the agglomerates will hydrate of course, but they are not in contact with the fluidity and gelation of the suspension. As these are released, they are free to participate in the set. The conclusion is that plaster is sensitive to the mixing intensity. By working fast, the mold can be poured.

Vacuum de-airing the plaster removes bubbles and improves its uniformity. De-airing was discussed in Chapter IV. Plaster slurries act just like ceramic slurries in de-airing. There is little problem in high intensity mixing and vacuum de-airing, but the mold has to be ready for pouring as there isn't time for procrastination. Usually, the mold is vibrated during pouring to remove air bubbles trapped on the walls of the form.

Marchant, McAlpin, and Stangle describe a process where the plaster slurry is indefinitely stored hot.[9] In this technique, there is no rush to pour the mold, and this would permit batching larger quantities of slurry that have to be stored hot.

Mold Preparation

The primary consideration for a mold is of course the geometry of the part and allowance must be made for shrinkage caused drying and firing. A pattern is made around which and into which the plaster is poured. Materials used for patterns should be impervious: aluminum, plastics, machinable composite, or well-sealed wood. Some pattern shops use a wood fiber composite called Ren Shape. Some pattern makers have computer-controlled, contouring capability so that any shape can be made. The two other parts of the assembly for simple molds are the base plate and the shell. A few types of shapes will be illustrated.

Right Circular Cylinder. The exterior shell can be anything with the right size and shape. A piece of plastic tubing cut off square and split down the side is useful. The base plate can be glass unless it is necessary to fasten the core to the plate with a bolt, in which case a polished chrome-plated, metal sheet of the type used to finish photographic prints (Ferrotype) is useful. Figure 6.29 shows a typical setup.

Here, the core is a cylinder bolted to the base plate. The shell is a length of plastic tubing split down the sides. Masking tape helps seal the splits. The bottom edge of the shell has to be sealed or the plaster will run out. One can use a fillet of modeling clay from a toy store to seal the base, but do not use the type that dries out. When using a glass base plate, the core can be cemented in place with a dot of hot melt adhesive or even rubber cement; this is a little risky as it might not hold. Include in the core the threaded hole that is useful for later pulling out the core from the plaster. Depending upon the length, one can use this type of mold for casting crucibles or tubes.

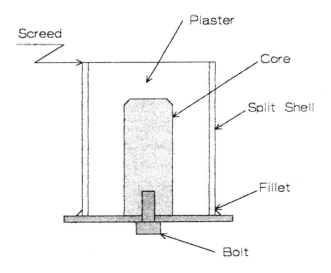

Figure 6.29: Plaster Form, Cylinder. A simple mold for making crucible or tubular shapes.

All of the surfaces in contact with the plaster are coated with a release agent, such as a mold soap solution of lithium stearate. Foundry supply houses carry a variety of products useful in mold preparation, such as release agents.

When the form is set up, the plaster is poured into the mold down one side to prevent entrapment of air bubbles. Overfill just a little bit. After a short time, the plaster surface will roughen due to formation of the gel structure. It is pasty and the top can be screed off flat. Setting is exothermic and the shell will become warm to the touch. After it cools down in a couple of hours, the shell can be removed and the plaster mold removed from the base plate. Put the bolt back into the core and pull it out. Plaster expands a little bit when curing, and this helps to remove the core that will be free.

It is important to clean up right away. Excess fluid plaster can be scraped into a container and left to harden; this is harmless and can go into the trash. Washing of the tooling requires a good flow of water in order to prevent the drain from clogging. A sump is a good idea for a ceramics lab because it will permit the solids to settle out.

Plaster has salts in solution that will migrate to the drying surface. This is not desirable on the casting surface, nor are finger prints or other irregularities. To prevent this, dry the plaster mold from the outside surface. This can be as simple as placing the mold upside down on the bench. Dry the mold the same way after slip casting to avoid accumulation of binders or surfactants on the casting surface. It takes about three days to air-dry a new mold. One can tell because it will feel warm rather than cold. The corners of the mold should be chamfered and scribed with an identification number.

Flat Plates. A flat ceramic plate can be cast on a plaster block with side restraints, but it will warp when fired due to density gradients across the thickness. A better design is to cast from both sides with the split along the diagonal.

It would be very difficult to remove a plate from the mold when the part is locked into a depression. By splitting the plaster mold along the diagonal, the plate is not physically restrained for removal. Another problem is that there is a large amount of contact area between a plate and the mold, but there are ways of handling this that will be described in a later section. There is also pipe along the center line. Ways to handle this will be discussed later.

Solid Plate. Two molds for making a solid plate are shown in Figure 6.30. The shapes of the molds are seen in the photograph. Note the separation along the body diagonal used for recovering the part.

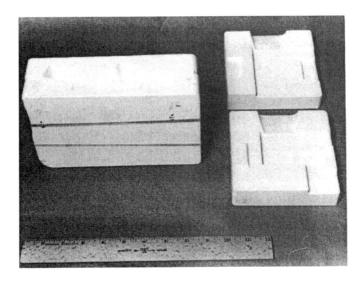

Figure 6.30: Flat Plate Plaster Mold. The photograph shows the mold configuration with separation along the body diagonal.

Solid disc.   The mold is shown in the next two figures. Because of the shape, it is easier to cast one edge as a disc and green machine the other later. The tooling and procedure for making the mold are a challenge.

The mold in the figures can be made from wood patterns except the shape form that is aluminum. Aluminum is easier to obtain release. The wood parts can be coated with a sanding sealer as long as the wood is porous, sanded, and then coated with three coats of gloss polyurethane. If the wood is not sealed well enough during the plaster cure, it can hydrate to some extent. The plaster mold is made in two steps: first make the lower part of the plaster mold and then cast the upper half on top of the lower half. By casting the two parts against each other, a perfect fit is assured. A coating of mold soap prevents the two halves from adhering to each other and to the wooden pattern. The whole pattern consists of 14 wood parts and one aluminum part. Let us walk through the mold making process to learn more about the craft.

Lower Half. Figure 6.31 is an assembly sketch of the lower pattern. Not shown is the box around the pattern that contains the plaster and makes the sides.

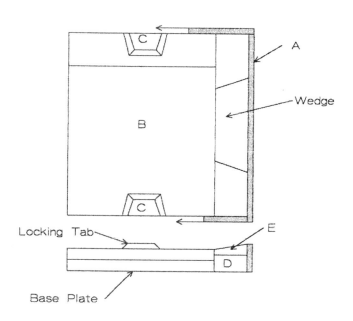

Figure 6.31: Mold Pattern, Lower Half. A box (A) surrounds the assembly into which the plaster is poured. The interior form consists of a base plate (B) and a half wedge. Locking tabs (C) will register the two halves together.

The box that encases the lower half is simply four slats of wood with the joints taped with masking tape and held with a rubber band. Figure 6.31 shows the lower half. These parts are marked on the figure and are as follows: lower form plate, two locking tabs, a back plate, and a half wedge. Instead of the locking tabs, one can use rubber balls. When the plaster is partially set, the balls are embedded half way into the plaster surface. The parts are assembled onto the lower form plate and held in place with rubber cement. Rubber cement does not form a strong bond, and the parts can be easily disassembled after use. Mold soap is lithium stearate paste diluted with water to a thin soap solution. It is painted onto the pattern and allowed to dry. Plaster is then poured over the pattern with a little bit just above the top. When set to a stiff mud, the excess is screed off, and the plaster is allowed to cure.

Upper Half. The upper part of the pattern consists of the shape form, the box extensions, and the full wedge. After removing the lower half pattern parts, clean the plaster by scraping off extraneous material. Do not touch the surface that will become a casting surface. The lower box is reassembled around the lower half and the upper half parts are put into place. Figure 6.32 depicts such an assembly.

A full wedge replaces the half wedge. Upper box parts are simply extensions of the box to accommodate the upper plaster. Aluminum was chosen for the form because of its dimensional stability and easy release from the plaster. After taping the joints again, coat everything with mold soap and allow it to dry. Plaster is poured over the lower half and cured. When disassembled, cleaned, and dried, the plaster mold is ready for use.

A few things should be brought to attention. Since the two halves were cast against one another they fit perfectly and the shape of the cavity is faithfully produced. The locking tabs assure that the two halves register perfectly. The wedge forms a slip reservoir, and the mold separates along the body diagonal.

Sometimes, it is necessary to make a split mold as this could be the only way to remove the part. Procedures for these molds are essentially similar to those shown here where the mold parts are cast against one another with some kind of interlocking tabs. Complex shapes can be made this way. Since it is labor intensive and the plaster has a limited life, one should consider other forming methods. In the plant, the working molds are made from master case molds and the process is not so labor intensive.

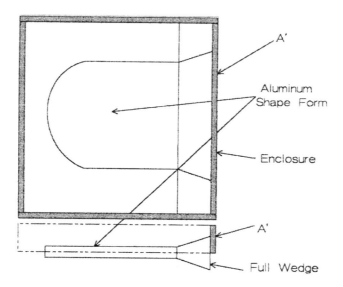

Figure 6.32: Mold Pattern, Upper Half. The plaster is poured into the box (A') onto an aluminum plate that forms the casting cavity. A full wedge replaces the half wedge to form a slip reservoir.

Air Release. Production molds sometimes have cloth tubes embedded in the plaster that can be filled with compressed air. A quick disconnect is cast into the plaster for introducing compressed air that breaks the cast part free. Figure 6.33 is a schematic representation.

When making the form, a wire frame is placed in the assembly and anchored to the pattern. Cloth tubes are tied to this wire frame and connected to a quick disconnect, which is exterior to the assembly. Plaster is poured over the tubes so that they are embedded in the plaster.

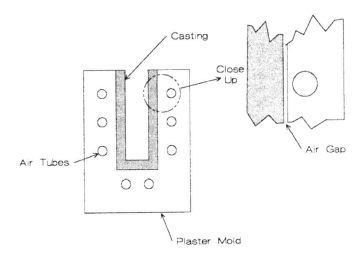

Figure 6.33: Air Tube Release. Fabric tubes are embedded in the plaster to provide release when they are pressurized.

Check List, Plaster

- Use #1 pottery plaster.
- Store plaster in a closed container.
- Mix the plaster in one of three ways: traditional, high intensity, or hot.
- Vacuum de-air.
- Pour into forms coated with a release film and vibrate.
- Forms can be metal, plastic, sealed wood, or a composite material.
- Air release can be built into the mold if needed.
- Screen off the excess and trim the mold.
- Slowly dry from the outside surface.

Slip Casting

Earlier discussions on preparing slip, storing slip, and making plaster molds lead to the actual casting process itself. Slip casting is an old craft and is still extensively used for sanitary ware and technical ceramics. Often, it is the preferred manufacturing method.

Casting Mechanics

Casting proceeds by pouring the slip into the plaster mold, usually down one side to prevent incorporation of bubbles. Bottom filling is a very good technique, but lab setups are usually not this complex. Immediately upon contact, a slip cast wall is formed. This wall plays a major role in the rest of the casting mechanics. Permeability and capillary pressure of the plaster ordinarily do not dictate casting as long as they are sufficient. Since the initial cast wall is less permeable than the plaster (for fine-grained slips), it decides the casting mechanics and is the source of the rate-limiting step.

The amount of water in the plaster mold establishes the initial casting conditions. When the plaster is too dry, the cast forms entirely too fast and is irregular. When the slip is too deflocculated, the initial cast is as hard as a rock and for all practical purposes the casting stops due to the low permeability. Removing the cast will be very difficult since there is negligible drying shrinkage. Drying shrinkage is necessary to loosen the cast part. Alternatively, one can use a split mold, which works part of the time. The plaster can be tested by placing a vertical tube filled with water on top of the mold and measuring the rate at which the water is adsorbed. When the plaster is too dry, pour some water into the mold, let it sit for about five to ten seconds, pour it out, and let the mold sit for a minute or so. If the mold is too wet, put it back into the oven at 60 °C. Deflocculated slips can be adjusted in many ways. Trial and error on surfactant, pH, and binder concentrations can experimentally adjust the slip to where it casts properly. Since the addition of the binder changes the rheology, there is no simple way to predict exactly how it is going to work. One can do a simple test. This test is to measure the casting rate of the slip on a plaster surface. Figure 6.34 is a sketch of such an apparatus.

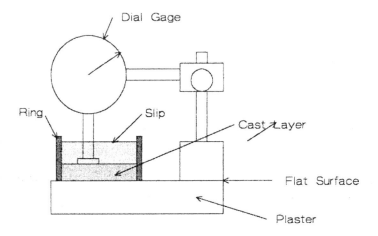

Figure 6.34: Casting Rate Device. A dial micrometer is used to measure the thickness of the cast layer with time.

The plaster can be cast on a pane of glass. Glass imparts a smooth and flat surface to the plaster needed as a base for the dial micrometer. The mold is just a short section that is cut from a plastic pipe and is sanded smooth. Keep the slip covered between measurements to keep it from drying. Measurements are timed, of course, and taken in a different place each time. Lift the micrometer and clean the tip after each measurement. Thickness versus time will plot as a parabola if it behaves as it should.

Coarse-grained slips with high-solids loading can cast faster depending on the PSD in the fines. The test shown in Figure 6.34 may need to be scaled up because of the large grain size.

After the measurements are completed, examine the cast part while it is still wet. Casts can be flabby, hard, or somewhere therein. The best consistency is somewhere therein but leaning toward the hard side. Dry the disc and fire it. Measure the drying and firing shrinkage and density. A linear shrinkage of a little less than 20% is achievable for fine-grained ceramics. Structural ceramics often have 100% of fired theoretical density as a goal. Coarse grained ceramics can have almost no shrinkage, and the density is usually between 65-90%, depending on the formulation. Lots of materials fall between these end members. Data from reverse engineering will help to guide the choice of formulation.

Mold Filling

It was common practice to place a ring on top of the mold to act as a slip reservoir. As casting proceeds, the slip is drawn down into the mold keeping the mold full of slip. To prevent drying, the reservoir is covered during casting. A watch glass or plastic film is useful for this.

In ordinary slip casting, there are two methods: drain casting and solid casting. In drain casting, the cast wall is built up to the desired thickness and then the remaining slip is poured out. One can measure the wall thickness with a probe. This method is not too accurate but is often good enough for a lab procedure. If more accuracy is needed, one can use appropriate micrometers for this type of measurement. Solid casting is a procedure where the slip is cast into a solid part, just like the name implies. One can locate the top of the cast with a probe. It is cast a little long and later trimmed off. To avoid distorting the part when trimming, always cut toward the plaster. Solid casting can also be used to make a hollow part such as a tube in which case a central core is placed in the mold. The core can be a casting surface such as plaster, in which case it should be collapsible or it will never be removable from the mold. Since the casting will shrink upon drying, it will crack if the core is left in. This makes timing an important consideration. Pull the core just when the cast is completed. One can also use rigid plastic foam cores. These are more flexible and put less stress on the casting. Polystyrene foam is very soluble in acetone and will collapse when squirted.

Pipe. Pipes are voids in the center of the casting that result from pockets of slip that are cut off from the slip supply. Entrapped slip will continue to shrink as it loses water, creating voids. Figure 6.35 shows such pipe voids.

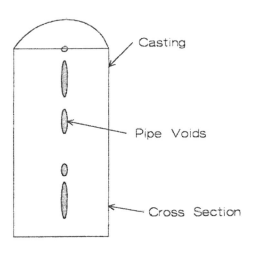

Figure 6.35: Pipe. The figure is shown cut in half. Pipes are the central voids.

Sometimes, pipes are not a problem as they do not extend to the surface. When it is a problem, there are a couple of ways to cope with it.

While the article is casting, the mold can be vibrated. An electrical, vibrating table is useful, with a fixture holding the mold on the table. Vibration intensity varies with a rheostat set at an intermediate value that causes enough vibration intensity to keep the slip fluid. Fine-grained slips are pseudo-plastic, so vibration makes sense. Coarse-grained slips have serious settling problems and vibration is not such a good idea. With the coarse grains nestled at the bottom and the fine grains on the top, the body will not shrink uniformly during firing, and will most likely crack. Settling can be slowed by increasing the slip viscosity. Coarse grains are not sensitive to surface chemistry as gravity dominates settling. An increase in viscosity will slow things. One can accomplish this by increasing the fines, flocculating the fines, or by adding an organic thickener. However, higher viscosities will also slow the casting rate that in turn increases settling time.

The second remedy is to use a tapered mold, illustrated in Figure 6.36.

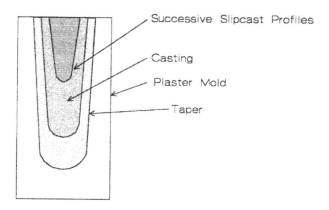

Figure 6.36: Tapered Mold. The taper avoids pipes as the central portion of the cast is always connected to the slip reservoir.

As the body casts, the taper prevents entrapment of air and a void-free solid cast results. Vibration is useful as the part casts; this helps to keep the slip fluid. When the cross section is square, fluid dynamics can draw air from the mold into the cast at the corners. Beveling the corners on the pattern helps to avoid this problem. Corner air bubbles are more likely when the plaster is too dry and the casting rate is high, in which case prewet the plaster with water to slow the casting rate. This tapered mold shape is probably not the geometry that is desired. Green machining, which will be described in Chapter VII, is a good way to reshape the part. In the lab, the priority is often to make a few parts for testing. The problem of producing the part will be addressed later. Production is inescapable if the research is successful. Different forming methods should not be overlooked should you get yourself into trouble making solid cast parts.

Long parts. When casting long parts, the slip at the bottom of the mold cavity is at a significantly higher pressure than the slip at the top. Hydrostatic pressure causes the bottom to cast at a higher rate than the top, producing a tapered wall. When solid casting, this is an asset to reduce pipe just as a tapered mold does. When drain casting (say a tube), a tapered wall may not be acceptable. The mold can be laid almost flat with vertical tubes at each end for the fill and for the riser. The riser end has to be higher than the fill end for air bubbles to escape. Coarse-grained slips will settle toward the bottom side of the cavity that it creates in homogeneity.

Drying Slip Cast Parts

It is usual to just air dry the part in the mold to the point where it is self supporting and can be removed. Keep in mind that water vapor is denser than air. One should tilt a drain cast tube in the mold at an angle to allow the water vapor to flow out. Turn the mold periodically for uniform drying. Large parts are more sensitive to drying stresses than small parts. Humidity drying is a very useful process especially for larger pieces, but this is not a lab process. Restricting the drying rate is an alternative. Since there should be some drying shrinkage, let the part equilibrate slowly in its moisture content to minimize stresses. Drying can also be done in an oven at 40-60 °C. It was mentioned earlier that solid cast parts are time sensitive. Experience will establish the right time to remove the part. Leaving it in too long can result in a reject. Always, set the part on a permeable or perforated base so that it dries uniformly. Microwave or dielectric drying has been around for a long time, but is not widely used. Depending upon absorption of the microwave energy in the part, heating can be superficial, creating temperature gradients or penetrating. Suppliers of this equipment can run tests.

Part Release

It is frustrating when the part will not release from the mold. There are several things that can be done to alleviate this situation.
- Drying shrinkage is necessary for the part to release from

the mold surface. Drying shrinkage is obtained by casting a slip that is just a little bit flocculated.

     - Split Molds are often used especially for platey shapes as shown in Figure 6.37.

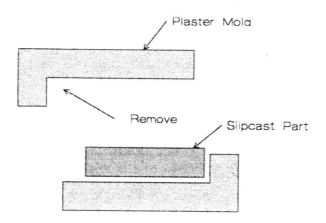

Figure 6.37: Split Mold. Molds separating along the body diagonal provide release of the cast part.

     This view is simplistic. A photograph of a split mold was shown earlier. The idea here is to design the mold so that the part is free to release. Here, the split is along the body diagonal. A cylindrical mold is split in half along its axis. The two halves have to fit together correctly as previously shown by using interlocks. In a ceramics lab, the molds are generally simple, but sometimes it is necessary to become more complicated.

Compressed air. A mold with air channels was shown earlier. One can use compressed air in other ways to help release of the part. One way is to place an air nozzle at the interface between the plaster and the casting. Air under pressure can break the part loose and release it. This is shown schematically in Figure 6.38.

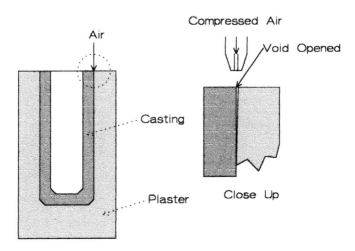

Figure 6.38: Air Release. Placing an air jet on the edge of the cast part can help to obtain release.

A better way is to drill a small hole in the bottom of the mold, intersecting the die cavity. When casting, the hole will fill with slip and cast. A piece of tape on the bottom of the hole will prevent the slip from running out at the start. Figure 6.39 depicts such a mold.

The part is forced out of the mold with air pressure that can cause it to become ballistic, so it is a good idea to catch it before it gets away. Also, turn the air pressure down just sufficient to break the part free.

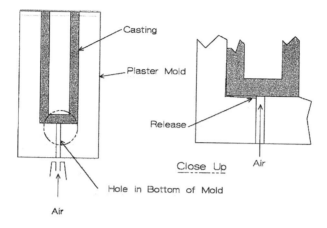

Figure 6.39: Bottom Hole in Mold. Compressed air entering the mold at the base pushes the casting out of the mold.

Corn starch. When there is adhesion between the plaster and the cast, a release film on the mold cavity can solve the problem. Pour a dilute water suspension of corn starch into the mold for three to five seconds before pouring it out. This leaves a thin layer of starch on the mold wall. Unless heated, starch will remain as particles, so the coating is a permeable film. This procedure will not address all of the mold release problems, but it can help to eliminate adhesion. When the part is fired, the starch burns off cleanly.

A suspension of starch and water is very dilatant in a highly loaded concentration. This is an easy demonstration of dilatant rheology.

Timing. Sometimes, there is an optimum casting time when the part should be demolded. There are two conditions: drying shrinkage and enough rigidity for the part to be self supporting. The part will not dry uniformly while it is still in the mold. The preferred technique is to release the part when possible consistent with rigidity. Use of a plaster core creates a compilation in that the casting will shrink onto the plaster core and put the part in tension, perhaps cracking it. One needs to have experience to determine the optimum demolding time for each case. There may be a problem with a system; however, one can get lucky and not have to worry excessively about timing.

Fuel Oil.[10] A thin coating of fuel oil rubbed onto the casting surface can sometimes prevent sticking. When all else fails, beat the bottom of the mold with a rubber mallet. Sometimes, this will jar the part loose, but often it will chip the plaster mold.

Check List, Slip Casting

- Adjust the plaster moisture.
- Measure casting rate and adjust flocculation.
- Pour the slip down the side.
- Determine demolding time.
- Rectify problems, refer to text.
- Measure drying and firing shrinkages.

7.0 RELATED CASTING PROCEDURES

There are other methods for consolidating slips into parts. Each has advantages and disadvantages.

Pressure Casting

Pressure casting and filtration casting are closely related. Perhaps the difference is that, in pressure casting, one uses plaster as the adsorbing medium while, in filtration casting, one uses a porous polymer as the medium or filter. One type of filtration casting has been patented and is commercially available. Unfortunately, it is necessary to obtain a license before using the method for experimental laboratory purposes. A license requirement is overkill for a lab when one does not know if the process is going to be commercialized.

Polymer molds are dedicated to a single shape from which many parts will be cast. Of course, plaster molds are also dedicated to one shape, but they can be easily made in the lab and are a lot cheaper. A sketch of a pressure casting apparatus using a plaster liner is shown in Figure 6.40.

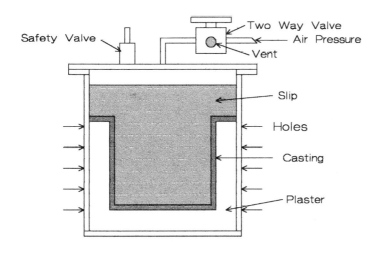

Figure 6.40: Pressure Casting Device. A metal shell has a plaster mold cast inside. Holes are drilled in the shell to provide escape of air. Always include a relief valve for safety.

Explanation: The shell is 0.25 inch thick steel with a welded flange. The lid is a steel disc with threaded holes for the air inlet, vent valve, and the pressure safety valve. Holes were drilled into the shell wall to provide removal of the water during casting. Arrows in the sketch point to these holes. An O-ring seal is at the junction between the lid and flange. Slip is poured into the plaster cavity at a level high enough to furnish a reservoir during casting. One can select the pressure, but 60 psi is prudent and adequate for this design. Casting is substantially faster by using pressure. For example, a 0.25 inch wall for sub-micrometer alumina was built up in about 0.5 hours. More importantly, one can make parts with pressure casting, but this may not be easily feasible by conventional slip casting. Pressure casting can be enabling! One part that can be made is a flanged jar with fired dimensions of 13-cm diameter flanges and 13-cm height. The slip to make this part can have a d_{50} as small as 0.6 μm.

When thinking about coarse-grained slips where settling is a serious problem, pressure casting can greatly reduce casting time and by that alleviate this problem. One can design apparatus that is not too expensive or difficult to use. Pressure casting deserves much more attention than it is receiving. Let us speculate a little.

Possible Pressure Casting Mold Designs

PVC has tensile strengths as high as 6000-10,000 psi, far more than needed. A length of PVC pipe can be fitted with standard end flanges and plates for top and bottom closures. One can either use drain tubing or use a solid tubing with holes drilled in the sides. Line the tube with a filter that could be plaster. A sketch is shown in Figure 6.41.

Clamp the assembly into place on a casting bench. The casing has holes so that the water can escape. Inside is a layer of a permeable material that either absorbs the water or acts as a filter, but it retains the solid particles. Plaster would be a good place to start. An impervious sleeve at the top prevents casting in that location so that a slip reservoir is maintained. One can pour or pump the slip into the cavity. When pumped, the device does not have air pockets. One can control the pressure by a sensor that activates the pump. As the slip is pressurized, it deposits a layer on the wall that builds up in thickness with time. The casting rate increases proportionally with the applied pressure, resulting in a shortened casting time. Once the desired wall thickness is obtained, the pressure is released, the toggle clamps are sprung, and the mold and top are removed from the casting bench. Now, the excess slip is drained out and the part partially dried by setting the mold on an air supply that pushes a flow of air through the casting. The same air supply, by increasing the pressure, can pop the casting out of the mold.

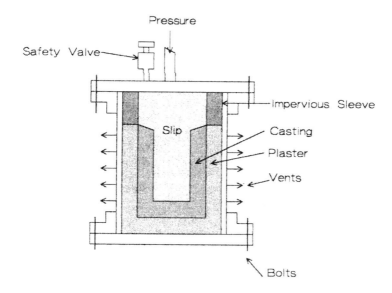

Figure 6.41: Pressure Casting Tooling, Speculative. The tooling should be durable and inexpensive.

This above example is for a simple cylindrical shape. There is no reason why one cannot fabricate more complex shapes in the same manner, though this may require a segmented permeable liner to provide release of the part.

Figure 6.41 would have size limitations due to the availability of PVC pipe sizes and the greater stresses on larger sizes. Let us speculate again. Figure 6.42 is a schematic of a hypothetical casting device for a large part, say two feet in diameter.

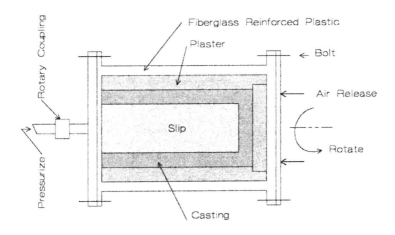

Figure 6.42: Large Rotary Slip Casting Device, Speculative. The tooling should prevent settling and increase the casting rate.

The shell is fiberglass-reinforced plastic with perforations. End closures are bolted on. Rotation on a horizontal axis at one or two rpm frustrates settling of the coarse particles. The sketch shows the mold filled with slip under pressure. It is very difficult to pump a coarse-grained slip as it will segregate in the pump and supply lines. It would be better to tip the assembly up to a vertical position and pour the slip in directly from the mixer. Then, immediately seal and place the assembly horizontally with rotation.

It is very likely that the plaster will crack when the slip is pressurized. This does not matter, as the slip will fill the cracks and seal them. One can release the large part with compressed air, but again the assembly should be tilted vertical so the part can slip out onto a support structure. Release will be easier if the plaster mold has a slight taper so that, as a minute amount of release is obtained, the part will slip right out. Since the piece is large and heavy, the assembly is turned upside down and the mold lifted off the casting. Remember, this has not been done and is only a speculation. One can visualize all kinds of things that could go wrong.

All of these speculations along with a little experience is convincing enough that pressure casting is a superior way to cast ceramics and should receive more attention.

T Casting

T casting is an abbreviation for thixotropic but this can often be a misnomer. Slips used in T casting are coarse grained and are often pseudo-plastic with a high-solids loading. For true T casting, one can use a variety of thixotropic additives. These range from chemically modified clays, hectorite, fumed silica, and organic compounds. Thixotropy often takes an appreciable time to develop in a slip. Clay-based slips can take a few hours for the viscosity to build up. One mostly uses T slips for refractory parts. They have a high-solids loading thus making them dilatant, but this is not desirable from a mixing point of view. One common T casting operation is casting saggers. The slip is kept moving in a mixer and then poured into a vibrating plaster mold that has both an outer body and a core. The sagger is cast bottom side up as shown in Figure 6.43.

The mold is made of plaster with two parts: an outer lining in the wood form and the core that has ring bolts for pulling the core out of the casting. Poplar is a wood of choice for wet applications. One can lacquer the wood with a water-resistant coating such as polyurethane and coat it with a release agent. Since the slip has high-solids loading, removal of just a small amount of water will solidify the part. Vibration lowers the viscosity of the slip when it is pseudoplastic and helps to fill the mold cavity. The base of the sagger is screed off flat. When one stops the vibration, the viscosity increases and the part casts with little settling. It is often convenient to design the wooden case so that it can be dismantled. As this mold has a plaster core, the demolding time is critical because the part shrinks and could crack. Do not leave until the core is pulled and the part demolded. Once the core starts to come out, the taper causes it to release easily.

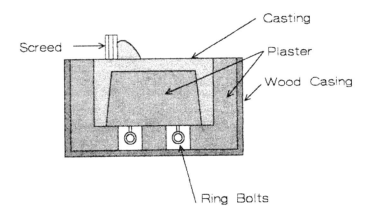

Figure 6.43: T Cast tooling for a Sagger.

When there is a true thixotropic mix, the slip will thicken and retard settling of the coarse grain. An important characteristic of T casting is if the viscosity increases quickly after pouring.

Though T casting is a little more labor intensive than other casting methods, but it is still useful.

Centrifugal Casting

Centrifugal casting has been used for a long time in the metals and plastics industries. There is some literature in the ceramics industry.[11] The advantages are that it can consolidate a part quickly, can make many parts in one cast, and can make intricate shapes with an appropriate mold. The disadvantages include keeping the apparatus balanced, cost of the equipment, and particle segregation. Segregation is a serious problem even with fine-grained slips. This can be approached by flocculating the slip. This is a good approach, but there still are segregation problems.

A centrifuge was used in some experiments in our lab. A plaster base was cast to absorb the water in the slip. After drying the plaster, the slip was poured into the centrifuge tube and was centrifugally cast. Casting was fast due to the high g forces. It took longer to get the centrifuge up to the required speed and then down again than it took the slip to cast. The centrifuge had a fixed rotor as seen in Figure 6.44.

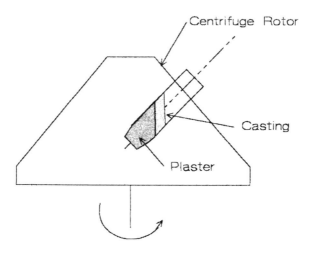

Figure 6.44: Centrifically Casting Experimental Apparatus

Due to the rotor geometry, the casting was on a slant as seen in Figure 6.44. The material is a slip containing an alumina powder with a d_{50} of 0.6 μm. It is segregated with fine material on top, which cracked on drying. In these particular tests, flocculation of the particles did not alleviate the segregation problem. In retrospect, a high speed centrifuge at 10 to 20,000 g was used. One should try a lower speed to prevent crushing the soft floccs,.

Gel Casting

One gel casting process is based on a polymer/solvent system with a catalyst to kick off the reaction.[12] Gel casting goes back in time in ceramics with cements, castables, zirconylchloride gels, alginate gels, and a variety of other systems.

Viscosity of fine-grained slips increases rapidly above about 50 volume % solids. To have a fluid slip, the solid's content is usually about 45%. Volume % solids and fine-particle size exacerbate the problems of drying and firing shrinkages, leading to distortions, cracking, and geometrical part limitations. Recently, considerable progress has been made despite these problems.

Complex parts have been cast with this process. When working with this technology, look for drying and cracking problems. The difficulties are inherent with fine-grained slips, but these are tractable. One approach to alleviating problems is by adding some coarser (4-5 μm) particles to the slip, thus increasing the solid content.

When the material is coarse grained with a variety of finer sizes, the permeability and the solid's volume % are both higher and drying problems are much less troublesome. Cements and castables avoid some difficulties with gels in that they are heavily loaded with ceramic particles that minimize drying shrinkage. While these do use a gelation process, they are not generally thought of as gel casting. The cementatious phase is calcium aluminate that hydrates and gels. Upon firing, the phase crystallizes. Calcium aluminate is a refractory material in its own right and is suitable as a bond in refractories. Additives are available for either shortening or lengthening the set time. A few grades are commercially available.

An alginate derivative with the trade name *Superloid* has been used for gel casting. A slip containing 1% or 2% Superloid will gel when immersed in a calcium chloride solution. Drying problems limit the size of parts made in this way. The high viscosity of the Superloid solution causes relatively low solids content in the slip and the parts will not have high strengths. Also, non-leachable Ca ++ remains in the body, changing its chemistry.

Injection Molding

Conventional injection molding uses either a wax or a polymer melted to impart fluidity. Complex shapes can be made by injection molding, often making this technique the preferred process. Lots of ceramic parts are injection molded, attesting to the usefulness of the process. This subject is voluminous, complex, and beyond the scope of this book.

A few comments appropriately point out some differences between polymer injection molding and injection molding of ceramics.

Polymer molding has additives that impart particular properties to the molded part. These can be fillers, antioxidants, opacifers, flame retardants, and pigments. These materials are added at relatively low percentages as compared to ceramic formulations. Ceramic injection molding is distinctly different because the polymer or wax is fugitive and only assists in forming. This makes for a substantial difference in formulation. Ceramic mixes are much heavier loaded, usually around 45-50% solids by volume. This leads to a different set of problems. Ceramic particles can segregate in mixing and in transport during processing. Design of the apparatus, molds, and procedures have their own criteria. Conditions in the die cavity are not uniform principally because of pressure and temperature gradients. When this leads to density gradients in the green ceramic, the part will distort or crack during firing. Data on strength is usually lower than that from other forming processes. An additional problem is binder burnout. Polymers swell upon heating and the part is about half polymers by volume. Polymers also create voluminous gas, during burnout, that require slow and controlled firing schedules. Since burnout involves reheating the part, it can slump as the binder softens. Large parts are saggered and supported by an adsorbent pack during burnout, introducing a separate step in the process. We have to keep in mind that ceramic parts are being made every day by injection molding so these problems are not intractable.

There is a very interesting paper that reveals segregation of phases during powder injection molding, largely due to design of the runners and gates.[13] Wherever there is a sharp turn or nook, the powder particles tend to pack to a higher volume % as in Figure 6.45.

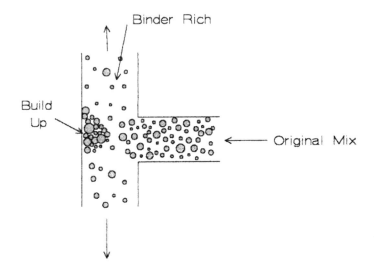

Figure 6.45: Solids pile up on a junction during powder injection molding.

The illustration could have as well been of a sharp corner. The consequence is that the mix is being segregated into high and low volume fractions of solids, which will remain after the binder is burned out. One example is where the location of the gate in a fired part has small cracks, probably because this was a solids-deficient area in the molding. The author of this paper illustrated the proper way to design dies for powder injection molding, as is redrawn schematically in Figure 6.46. The design avoids corners where segregation takes place.

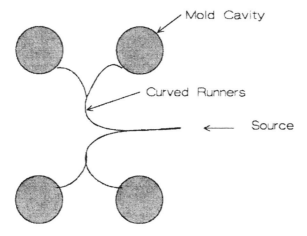

Figure 6.46: Runner design to prevent pile up of particles during powder injection molding.

Adiabatic Molding of Ceramics

So far, experimental processes have not been included in this text, except for one gel casting method. Adiabatic molding is an experimental process.[14] It is included because it has attributes that may offer some advantages.

This process is a water-based system. Water is an unusual material in that the liquid has a higher density than the solid. Because ice floats, pressure, rather than temperature, can be used to melt ice. This requires an explanation. Figure 6.47 is the phase diagram of water.[15]

In the example, at -10 °C at point A, when the pressure is increased to about 180 MPa at point B, the liquidus is crossed and the ice melts. The conclusion being that pressure can be used to melt ice instead of the usual way for melting by raising the temperature. Consider an example. Alumina with a median particle size of 0.6 μm can be mixed by hand with a 3%

PVA solution to make a dough. A pellet of this dough is placed in a die and put into a freezer. The die is configured to a crucible shape. Overnight, the assembly is equilibrated at -10 °C. As pressure is applied to the pellet the next day, the water in the pellet is melted and the mix then is injected into the die. When the pressure is released, the water solidifies and the part can be ejected from the mold. Dried and fired the crucible shape is produced.

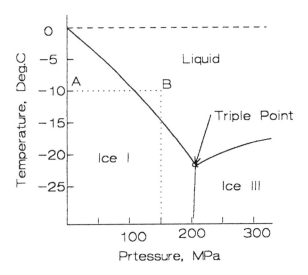

Figure 6.47: Phase Diagram, Water. Liquid water can be present above the triple point. Water frozen at (A) will melt when pressurized to point (B).

Figure 6.48 is an alumina crucible shape with the dimensions; 16 mm OD, 14 mm ID, 30 mm long, and 21 mm deep inside. This is not an easy part to make by other methods as it has a 1mm wall over a 9 mm thick base.

The process works as expected with pressure being effective to melt and then freezing the mix, but there is a problem. When the pressure is released, the water freezes and expands. Since this is a cylindrical mold, the part was free to expand only along the axial direction. This results in shear within the piece, resulting in shear cracking. The problem is not intractable as articulated tooling could provide the space necessary for the part to expand freely when the pressure is released. This is speculative now as this could be a little tricky. Internal pressure and temperature in the die are uniform after the dough has been injected. Therefore, mix viscosity is uniform throughout the die cavity. In turn, packing is uniform avoiding some problems inherent in conventional injection molding. Another good feature is that the dough is mixed at room temperature. This results in a useful process simplification.

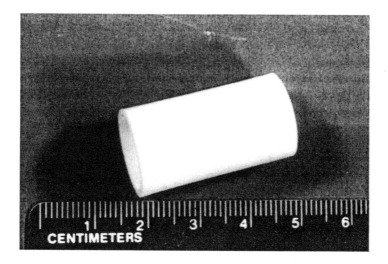

Figure 6.48: Adiabatic Molded Crucible. The shape has a large discrepancy between wall thickness and base thickness.

Check List, Other Casting Procedures

- Pressure casting: fast cast, thick walls, reduce settling, safety valve
- T casting: coarse slips, high solids, timing
- Centrifugal casting: segregation, flocculate slip

- Gel casting: warping, cracking, coarsen slip, high solids
- Injection molding: density gradients, low solids content, expensive tooling for a lab process, preventing segregation of solids by design.
- Adiabatic molding: experimental, uniform high density, expensive articulated tooling for a lab process

8.0 EXTRUSION

Equipment for extrusion was described in Chapter V. Extrusion is a common forming method for making cylindrical shapes. A cylinder is being used in its broader sense, where it is any three-dimensional form consisting of parallel straight lines. The cross section can be square, rectangular, oval, a star, or a honeycomb, to name a few cylinders. Common extrusions are bricks, tubes, solid cylinders, mill lining blocks, and honeycombs used as heat exchangers and automobile exhaust catalyst supports.

Preparation of the mix was described in Chapter V. It has to be plastic like stiff mud. It is usually flocculated. Clay bodies are naturally plastic especially when they contain ball clay. A small amount of clay or other alumina silicate such as Hectorite can be added to oxides or carbides to increase fluidity. Organic, long-chain polymers or alginates can also be added to increase fluidity.

The extrusion is run out of the die onto a table and wire cut to length. When the extruder is vertical, it is run out onto a board, cut, and swung away horizontally. In both cases, the idea is to keep the extrusion straight. If it is bent, the particles on the compression side will be compacted, while those on the tensile side will be extended. Subsequently, the part will warp when fired even if it is straightened back out. Common practice is to use V blocks to keep rods and tubes straight as they dry.

There are many variations on extrusion; two are briefly described below.

Honeycombed. The extrusion mix in one instance contained methyl cellulose.[16] This material is unusual as a water solution of it becomes more viscous when heated. To preserve the shape and keep it from sagging, the extrusion die is heated causing the shape to stiffen.

End Flange. A plate with a cavity can be placed on the extrusion die, forming a flange on the end of the piece. The plate is then removed and the length extruded.

9.0 DRYING PARTS

Green ceramic parts are dried after forming, provided of course that a liquid is present. Drying increases the green strength and readies the part for sintering. When the interstitial phase is a polymer or wax, such as in injection molding, the removal process is called *burnout*.

Drying Conditions

Conditions of the body that are important are thermal conductivity, permeability, and drying shrinkage.

Conditions in the dryer that are important are temperature and relative humidity.

Before these are discussed, consider what happens as a part is dried. Depending on particle packing, most parts will shrink when dried. The outside shrinks first. Since it is restrained from shrinking by the interior, it may crack as shown in Figure 6.49.

This is particularly troublesome when the part is thick and the interior may not even be warm. The problem then is to program drying so that differential shrinkage between the surface and interior does not exceed the tensile strength of the material.

In extreme cases, internal pressure can burst the part when the water vapor is being generated faster than it can diffuse out of the body. Fine-grained, dense bodies are susceptible to bursting when dried too fast.

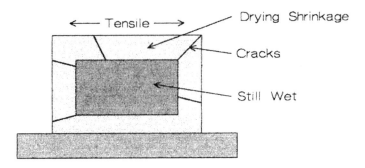

Figure 6.49: Drying Shrinkage. Uneven shrinkage can result in cracking.

Permeability

Water evaporates during drying and has to escape from the interior of the article. Remember that one mole of water (18 g) will create 22.4 liters of steam at STP. Since it is hot, there is even a greater volume of gas that has to escape. When permeability is low, internal pressures can build up and burst the part. Fine-grained ceramics have low permeability and have to be dried slowly. Laboratory practices are not especially urgent so a relatively low temperature, say 60 °C, is about right. A flow of air in the oven helps to remove the water vapor from the oven. Keep the vent open. Coarse-grained bodies have higher permeability and are not always thoroughly dried as they can dry in the initial ramp of the firing curve.

Setting

It is not uncommon to just air dry parts setting on a suitable surface, sometimes configured to match the piece. The setting of the wet part is often critical. When set on an impervious surface, the top and sides can dry but the bottom stays wet. One can lose parts this way by distortion and cracking. Always set the part on an absorbent or mesh structure so that air can reach all surfaces of the piece. This is also true for oven drying. A fine, expanded, metal mesh works well. Depending on the shape, one can turn the part from time to time to help attain even drying on all surfaces.

Binder Burnout

After drying, the binder remains in the body and is burned out during the sintering cycle. The initial heating ramp is slow (0.5 °C/min. for sensitive materials).

Binder burnout for injected molded parts is another story. The binders are thermoplastic and are present in large amounts. There has to be sufficient liquid present during molding to make the compound fluid.

Wax injected parts have a low viscosity fluid when they are reheated. The part will slump. To get around this, the part is saggered and packed in an absorbent material. Clay is sometimes used. Once a little of the wax is removed from the surface, the part could be self standing.

Injected molded parts with a polymer vehicle have a much more viscous liquid when the polymer melts. It is not common practice to sagger and pack these materials. When the polymer is heated, it expands, and then gives off a large volume of gas. Since the fluid is viscous, the escape of the decomposition gases is difficult. However, this is done in large volumes every day, indicating that the problems are solvable.

Types of Dryers

Ceramic parts can be dried in many ways, depending on the properties of the ceramic and the equipment available. A few types will be discussed.

Air Drying

For this, one just sets the part to be dried on the bench. This is often a suitable method. There are two problems with this method, however. First, the part has to be set on an open mesh or absorbent material so that it dries evenly. The second problem is that, if there is dust in the air, the part should be loosely covered.

Oven Drying

This is the most common laboratory drying method, using an electric oven. The oven has a temperature control, temperature indicator, expanded metal shelves, and a vent. The expanded metal shelves allow drying to take place on the bottom of the part. Do not set parts on an impervious shelf as they will not dry evenly. It is good practice to place thermometers at different places in the oven to see how much variation occurs; there will be some. One also has to think about air circulation in the oven. Usually, the heaters are on the bottom so that the heated air can rise through the setting. The temperature control is often at the top of the oven some distance from the heating element. If an air blockage in the oven occurs because it was overloaded or the shelving blocks air flow, the heater will overheat, ruining the ware and possibly ruining the oven.

Some ovens have a blower or fan to circulate the hot air. This is a good practice.

Humidity Driers

Humidity driers control both the relative humidity and the temperature. By controlling both, the rate of drying of the parts is controlled. Humidity is gradually reduced as the parts dry. Temperature can also be controlled along with humidity. This makes humidity drying the preferred method if parts warp or crack when dried by other methods. The equipment is more expensive and not usually seen in the lab. This is partially also true because lab parts are generally small. Large parts that have low permeability are candidates for humidity drying.

Microwave/Dielectric Driers

With this method, the part is heated more uniformly as the microwave energy penetrates the body and couples with water thus increasing the temperature. Since heating is more uniform, drying tends to be more uniform. There are two snags. As microwave energy is adsorbed by water, the intensity of the field diminishes. For a thick part, the center receives less energy than the exterior. The second snag is that the water vapor still has to escape or the part could burst. Microwave drying has not been popular for general laboratory use. It is popular for drying films or fibers on continuous production lines.

REFERENCES

1.   Metals Handbook Ninth Edition. Am. Soc. for Metals, Metals Park, Ohio.
2.   *Bulletin of the Amer. Ceram. Soc.*, Editorial and Subscription Offices, P.O. Box 6136, Westerville, OH 43086- 6136.
3.   W. Kingery, Ceramic Fabrication Processes. Cambridge, MA: Cambridge Technology Press of Massachusetts Institute of Technology,1958.
4.   Robert A. Thompson, "Mechanics of Powder Pressing: I, Model for Powder Densification," *Bull. Am. Ceram. Soc.*, 60[2]237-243(1981).
5.   Robert A. Thompson, "Mechanics of Powder Pressing: II, Finite-Element Analysis of End Capping in Pressed Green Powders," *Bull. Am. Ceram. Soc.*,60[2]244-251(1981).
6.   Alan G. King, Method for Isostatic Pressing of Formed Powder, Porous Powder Compact, and Composite Intermediates. U.S. Patent 5,244,623 (1993).
7.   E. L. Whiteside, "Quality Control in the Plaster Mold Shop," *Bull. Am. Ceram. Soc.*, 45[11]1022-1026, (1966).
8.   Carl Pletcher, Personal Communication.
9.   David D. Marchant, James L. McAlpin, Tim Stangle, "Investigating Plaster Properties," *Ceramic Industry*, December(1994).
10. J. T. Jones, Personal Communication.
11. Frederick F. Lange, U.S. Patent 4,624,808 (1986).

12.   Ogbemi O. Omatete, Alan Blier, C. G. Westmoreland, et al, "Gel Casting Alumina," *Ceram. Eng. Sci. Proc.*,1991, 12[9-10] 2084-90.

13.   Karl Frank Hens, "Advanced Tooling and Molding for PIM," 1995 International Powder Injection Molding Symposium, July 19-21, 1995, State College, PA.

14.   Alan G. King, S. T. Keswani, "Adiabatic Molding of Ceramics," *Bull. Am. Ceram. Soc.*,73 [9] (1994).

15.   C.E. Weir, from P.W. Bridgman data, "System H$_2$O PT diagram Between 0 and 10,000 bar"; Fig.1915 in Phase Diagrams for Ceramists, Edited by M.K. Reser, *Bull. Am. Ceram. Soc.* (1964). Redrawn by the authors.

16.   F. A. Cantalope, R.I. Frost, L.M. Holleran, "Method for extruding thin walled honeycomb structures," U.S. Patent 3,919,384 (1975).

7

Green Machining

1.0 ADVANTAGES

Green machining separates the two steps of consolidation and shaping. For example, a lump of ceramic can be isopressed and then green machined to shape. There are several advantages to green machining: one can obtain a uniform high density packing; one does not require expensive tooling; fixturing is not time-consuming; and one can shape the part to precise dimensions. Fixturing is prevalent in crafts such as wood working, carpentry, and metal working. These are good sources for information.

Methods used depend upon the physical character of the body. A hard and brittle body is much more susceptible to fracture than a soft, compliant body. Fine-grained ceramics make for hard, brittle bodies, especially when they are compressed to a dense, green structure. Here, grinding works better than single point turning, where fracture is a serious problem.

Binders

Binders can greatly affect how the part machines. Strength and plasticity can be adjusted by the kind and amount of binder. PVA and acrylic emulsions are strong binders. Recently, Rhome and Haus introduced a new acrylic binder with very high strengths. They claim that parts can be machined with single point tools, but they hesitate when discussing drilling. Comparative green strengths are shown from their data in Table 7.1.

Table 7.1: Comparative Green Strengths

Binder	Strength (Mpa)
Acrylic	6.5
PVA	0.7
PEG	0.4

One commercial example of green machining is making spark plugs. A blank is isopressed and then ground by a form-grinding wheel. This process is very fast and produces an accurate and uniformly dense body.

2.0 LAPPING FIXTURES

An easy technique uses lapping fixtures for green machining a part. Design the fixture so that it has a large area in contact with the lap and is hard, thereby increasing its wear resistance. As the green ceramic is soft, it will lap to dimensions in a short time. Laps are often metallurgical types with an 8 inch D wheel and a Silicon Carbide paper on its surface.

Fixtures for Lapping Flats

Figure 7.1 depicts a typical fixture. The fixture has five parts: a base plate, two wear plates mounted on the base, and two thinner wear plates for lapping the obverse side flat and parallel. Flat-headed screws hold the wear plates on the base plate. Slotted holes in the wear plates provide lateral adjustment. The part is placed on the base and snugged up by the lateral adjustment. The fixture is held lightly on the lap when it is flooded with a coolant. (For water soluble binders, use kerosene as the coolant. Bisqued parts use water.) Let the fixture float on the lap as it is not necessary to use much pressure. Quickly, the green part will grind down to the plane of the wear plates. Now install the thinner wear plates, turn the part over, and repeat the process. Both sides are now flat and parallel. It is just like using a surface planer in wood working. With another set of wear plates, the edges can be squared up so that the part has all four sides flat, parallel, and square, just as would be done on a joiner in wood working. Next, square up the ends.

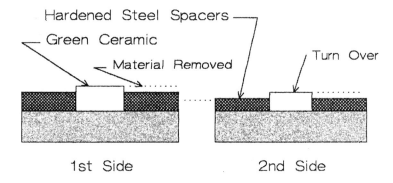

Figure 7.1: Lapping Fixture, Flats. Wear plates fix the depth of lapping. Different thickness plates are used to lap flat and parallel.

Lapping Ends

To square up the ends, use a different fixture, with the same principle but with a different shape, as in Figure 7.2.

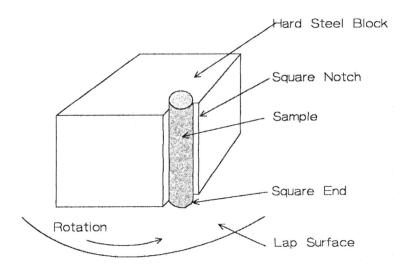

Figure 7.2: Lapping Fixture, Ends. A simple square block with a corner cutout is useful for lapping ends.

The fixture is simply a block of hardened steel with a notched corner. The part is held in the notch by hand with the fixture floated on the lap. Press the part to the lap with a finger. Direct the lap rotation so that it presses the part into the notch. Now, the other end can be lapped to length. If careful, tolerances of 0.001" are attainable. It turns out that this fixture found a major use in grinding; with metal-bonded diamond laps, the ends of glass thermal expansion specimens, which were lapped square and to length. Whenever the fixture wears out of square, it is remachined.

Chamfering

Often, one needs to chamfer the edges of the part with a fixture as shown in Figure 7.3. Two angle blocks that one can move to different spacings are fastened with set screws. Control the width of the chamfer by the spacing between the blocks. One can quickly grind the portion that protrudes. A similar technique was used to chamfer ceramic-cutting tools by holding the corner up against a diamond wheel. In this case, set the width of the land by the cross slide position of the grinding machine.

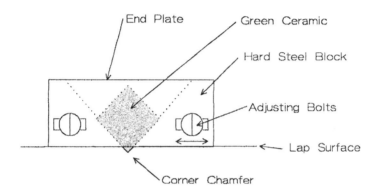

Figure 7.3: Chamfering Fixture. The chamfer is made by grinding away the exposed corner.

Holes

A good way to cut a hole in a part is with a core drill as shown schematically in Figure 7.4.

Core drills will be discussed in the following section on tools. Drill speed is dependant upon the diameter of the drill. The periphery speed is kept as a constant and, for this application, fairly low. One can also use metal tubing as a core drill with an abrasive grain and a coolant. Grain is fed into the cut during drilling. Chances are that the cut will not be as clean as that from a diamond drill, but if cut fat, the cut can be cleaned with a mounted point.

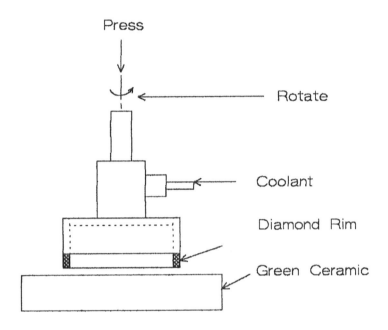

Figure 7.4: Core Drill. A core drill mounted in a drill press is used for green machining holes.

Cylinders

It is possible to green machine cylinders on a lap. Figure 7.5 is a fixture for lapping cylinders.

The fixture has two steel blocks that are adjustable up and down and locked in place to set the diameter of the cylinder. Fasten two hardened wear plates to the blocks with flat head screws. When the plates wear, they can easily be reground. The right block is on a slide that, with horizontal hand pressure, holds the part in place by way of centers ground into the ends of the part. The centers have to be in line, which can easily be done with temporary shims. One way to do this is to grind them by hand using an abrasive point in place of the rotating centers. With the ceramic mounted, the other hand rotates the part. Figure 7.6 depicts another idea for grinding cylinders.

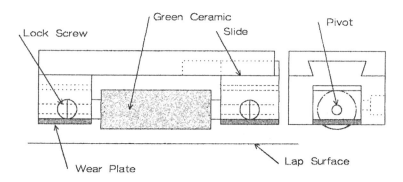

Figure 7.5: Lapping Fixture, Cylinder. The green part is turned by hand on a SiC-coated, abrasive lap.

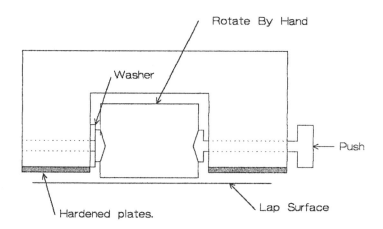

Figure 7.6: Lapping Fixture Alternate, Cylinder. Another design for green machining cylinders has more size flexibility.

This fixture is different because the thickness of the hardened steel facing sets the diameter of the cylinder. The facings are held in place with flat head machine screws. Again, hold the part with hand pressure by pushing on the right center, and rotate the part with the other hand as before.

3.0 TOOLING

Tooling falls into two classes: improvised tooling with fixtures and conventional machine tooling. Grinding is going to be emphasized.

Abrasives

Most common abrasives are bonded diamond and Silicon Carbide abrasives. One can use Silicon Carbide as a grain or as a bonded abrasive article, usually as a coated abrasive. When green machining, one uses an intermediate grit size. Since fine-grit, bonded abrasives tend to blind, they are ineffective. Coarse grit is too rough and can damage the part. The grit size range of 60 to about 120 is generally useful, but one can use a little finer grit size.

Coolants

One generally uses a coolant to wash away the swarf and keep the part from overheating. The coolant also washes the swarf out of the abrasive face, keeping it free to cut. Depending upon the binder, one can use water or Kerosene. Often, the binder will be water soluble, so this will exclude water as the lubricant. However, there is an alternative. The part can be bisque fired, and which water is the coolant of choice for grinding. The extra firing is not a major issue in a lab procedure, but it might not be cost effective in production.

The type of equipment will determine how the coolant is applied. If it is a lap or drip, one can use a squirt bottle. If it is on a machine tool, there will probably have to be a bucket and a circulating pump. Not all machine tools handle coolants, so one will need to improvise or try other equipment.

Kerosene is flammable, so it will require precautionary measures. Motors should be explosion-proof or better still, remote. The same is true with switches. If there is a fair amount of work, one will need to rig a vent. It is not a good idea to ingest this mist, nor to have its concentration built up to explosive concentrations. A small amount of Kerosene is not too bad on your hands. A large amount of Kerosene will dry and chap one's skin, so wear rubber or vinyl gloves. Never work without eye protection. Fire is the principal hazard.

Machine Tools

Machining operations generally are designed to make shaped surfaces such as planes, right circular cylinders, spheres, and cones. It is unlikely that special contouring equipment will be available.

Cutoff

One uses diamond cutoff saws for cutoffs. Diamond blades have the following characteristics: they have a continuous rim, are thin, and are metal bonded; probably with a high diamond concentration. For 3-6" diameter blades, the thickness should be no greater than about 0.030". Thicker blades put excessive stress on the part. Again, the grit size should be intermediate and not too fine.

A cutoff is done with a coolant to remove the swarf. Otherwise, the cut will gum up. For small work, a lapidary saw that dips the blade into the Kerosene coolant is not very precise, but can cut off the end of the part. Figure 7.7 is of a small cutoff saw that is suitable for small green machining cutoff work.

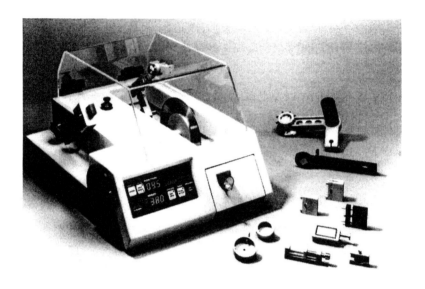

Figure 7.7: A cutoff machine. Diamond wheels are used to cut off the ends of parts or for cutting slots. (Courtesy of Buehler, Ltd.)

Larger equipment should be set up with a coolant supply and safety features. If equipped with a coolant capability, use a surface grinder with a cutoff blade. The same safety precautions, as above, will apply. Hold the part with a fixture on the bed. If it is flat, it can be blocked in with steel on the magnetic chuck. If it is round, one can use a V-shaped block. Near the end of the pass, the cut will break out, leaving a chip. Cut only part way through and then cut in from the other side to put the breakout in the center where it will be smaller and less obvious. One can also back up the part with a similar expendable piece to reduce any chip-out.

Surface grinders are an exception. Most cutoff on a green body is done off hand and the part trued up later. Wire saws are also available and have the advantage of not jamming the blade.

Laps

A lap with a Silicon Carbide paper is very useful for green machining. One can also use it with loose, abrasive grain. Apply Kerosene with a drip or squirt bottle. Hold the paper in place with a retaining ring or a pressure sensitive adhesive backing. When a loose paper starts to slip, wet the back side and it will stay in place. Most of these laps are 8 inches in diameter and have the choice of two speeds. They have a splash bowl and a drain that makes it easy to set up for the coolant. While designed for metallography, they are also useful for green machining of small parts. Grit size of the coated, abrasive paper is most useful in the coarser grits, from 60 to 120 grit.

Core Drills

These are abrasive tools used for drilling holes. A coolant is always used to carry off the swarf; introduce it down through the center of the drill. It is common to have these mounted on a drill press where there is a rig to take away the coolant. The working part is a metal-bonded, diamond-continuous- rim tool. Again, the grit size is moderately coarse. Let the drill do the work and do not push it too hard. There will be a breakout when the drill completes the cut. One can back up the piece by placing a blank scrap underneath for support to limit breakout, or drill in from both sides toward the center.

Form Grinding Tools

Grinding wheels can be configured to grind a contour. This is essentially a router using an abrasive cutter. A router is a versatile forming tool much used in wood working. One can use it as a joiner cutoff for making contoured edges, rabbets, dados, or to drill holes.

One creates the form during manufacture and trues it with a

diamond or crush-type, trueing device. As the grinding wheel rotates, the trueing device shapes the surface of the wheel to the desired contour. In use, the rotating part is fed into the rotating form grinder to make the desired shape. One configures spark plugs this way. In the lab, there usually are not enough parts needed to go through all this trouble. If a machine shop that has a tracing attachment and a tool post grinder attachment on its lathe can be located, then that might be a suitable way to go. Another alternative is to locate a pattern shop that has contouring capability, grinding and cutting tools.

Green machining equipment has been developed for larger scale operations as well. Figure 7.8 depicts a green machining center that uses three machines.

Figure 7.8:   Green Machining Center. Commercial green machining centers that feature abrasive cutting tools are available. (Courtesy of Chand Kare)

These are grinding machines that are CNC controlled, and each has nine axes for setups. With this much flexibility, coupled with lateral motion of the work piece, many shapes can be machined. This is somewhat elaborate for a standard ceramic lab; it all depends on the research requirements. Metal bonded mounted points of different contours are chucked in precision high speed spindles. Grinding is done under dry conditions in the sealed enclosure, and there is a facility for connection to a dust collector. The stresses and temperature of the work piece are affected by the speed control of the mounted point, depth of the cut, and the grit size of the grinding tool. This is more critical when grinding dry than when grinding wet.

Check List, Green Machining

- Grinding is the preferred method for green machining.
- Fixturing is a quick and accurate way to shape the part.
- Wear plates control the depth of cut.
- Coolants are needed to wash away the swarf.
- Kerosene is a preferred coolant for water soluble binders.
- Water is the preferred coolant for bisque-fired parts.
- Machine tools are available for green machining.

8

Firing

1.0 INTRODUCTION

Ceramics are sintered by raising them to a high temperature. This process consolidates the material, increases its strength, and usually causes it to shrink. This chapter includes sections on equipment, setting practices, firing procedures, hot pressing, and hipping.

2.0 EQUIPMENT

A variety of furnace designs are commercially available for all kinds of applications. It is better to purchase equipment than to build it yourself. In either case, there are design features that one should follow or look for.

Not all purchased furnaces are free from design or manufacturing errors. It is advisable to carefully check out the equipment before acquiring it. Rather than trying to be encyclopedic, the purpose of this section is to illuminate furnace design errors in the belief that one can avoid such errors in the future. Names of manufactures will not be included, as there have been many changes in furnace design and materials in the past few years.

Box Kilns

One can open these boxes from the front, the bottom, or the top.

Configuration of box kilns

Front. A problem with these kilns is that the configuration of the door is unstable with use. The steel supporting structure warps with use because of overheating and the door ceases to fit. This problem is progressively degenerative. The front of the setting will be at a lower temperature than the back. Backup insulting brick will also overheat, causing them to shrink. Bricks fall out. Another way to make a door is to pile bricks in the opening to seal it off. When the joints are staggered properly, the door will be okay, but it will not last. There are always odd shapes to fill the openings that are lost or worn out. And this can work, as it has for millennia.

Drop Hearth Kilns. Since access to the hearth is unobstructed, drop hearth kilns are very handy to load. However, if the setting is precarious, the piece may topple out when one opens the kiln after firing. Hearths can shake as they are moved up and down. Figure 8.1 depicts a lab size, electrically heated, drop hearth kiln.
There are several features about this design that are important. The size of the furnace is large enough to keep the ware away from the proximity of the heating elements. A 1-inch to 2-inch clearance is best. This ample furnace has a hearth about 16 inches by 16 inches. A large box also helps in air circulation, especially for an open setting. A ball-screw drive lifts the hearth. Mechanical lifts, such as a smooth ball screw, provide positive movement that results in little shaking. Refractories on the hearth are staggered so that there is no through line of sight for radiation. This is standard construction wherever there is an opening to the interior of the box. Raised hearths or fiber insulated hearths are necessary, as cold hearths produce uneven firing. In the kiln shown, the hearth plate has a shoulder on both the back and front. It is common practice to set pieces on a bed of

refractory grain, and one can use these shoulders as guides for screeding the grain bed flat. If needed, one can add a raised hearth set on blocks or vertical refractory tubes counter bored into the existing hearth stuffed with refractory fiber. Heating is with $MoSi_2$ elements that can be set to 1700 °C or higher. Higher temperatures are hard on both the elements and fiber refractories, but it can be done. Another appreciated feature is the expanded metal cover that is offset from the kiln shell. This is cool and prevents burns when the operator lurches against the furnace. Controls and gages are straightforward and can be programmed to just about any cycle, subject to the cooling rate maximum of the kiln.

Figure 8.1: Laboratory Furnace, Electric. Drop hearth kilns allow access to all sides of the setting. (Courtesy of CM Furnaces)

Bell kilns. Lift the box up to reveal the hearth. As before, there is access to the setting from all sides. Since the hearth is not moved, it is stable. The seal between the bell and the hearth is often a sand seal, at least in larger sizes. For some reason, bell kilns are not common in ceramic laboratories but are common kilns in plants.

Electrical Heating

There are few heating elements for an air atmosphere such as $MoSi_2$, metal, and SiC elements. Metal elements are very dependable but are limited to about 1200-1300 °C. This leaves us with two choices.

SiC. Use of these heating elements is wide spread and has been for a long time. SiC elements are good to about 1550 °C for general use in air. Under certain circumstances, they can be used to 1650 °C. A silica film that forms on the surface of the element inhibits the oxidation of the SiC. During heating and cooling, this film will crack. It will reform during the next heating cycle. However, if the next heating cycle does not reach a high enough temperature to form the film, the SiC will oxidize. Intermediate temperatures of about 600 °C and 1000 °C shorten the life of SiC. One should also remember that resistivity of these elements decreases as the temperature increases. Controls have to adapt to this condition and suitable controls are available.

Electrical contacts and connectors on the ends tend to oxidize, increasing resistance and requiring routine maintenance. There is a problem with the bank of resistors not being in balance. As these get older, the resistivity increases and usually when serviced the old elements are not matched to the newly replaced ones. Some elements run hot and some elements run cold. At worst, one will need to replace the whole bank, and this is expensive. It is important to follow the manufacturer's instructions on the design of the installation. One of these requirements is to leave about 2 inches or 3 inches between the resistor and both the wall and the ware. This helps to avoid hot spots and one is less likely to bump and break the heating element, which is brittle.

MoSi₂. These heaters are wide spread in use. They are good to at least 1600 °C for the regular type. One can extend the temperature to 1900 °C with the super type heaters, but this is pushing it. Use at high temperatures shortens the life of the element, but sometimes this is unavoidable. These elements are available in a variety of sizes and shapes. Figure 8.2 depicts some MoSi₂ heating elements.

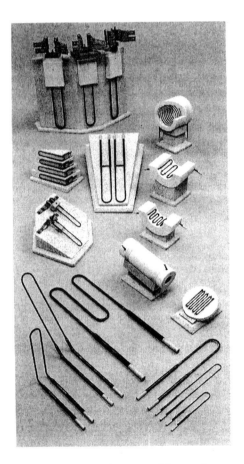

Figure 8.2: MoSi₂ Heating Elements. These elements are available in a variety of sizes and configurations. (Courtesy of Kanthal Corporation, TM)

Try to avoid the smaller diameter elements as they are fragile and warp. As with SiC, a silica film that forms during firing protects the element. When one does not follow the installation instructions, replacing elements can be expensive and the down time extravagant. $MoSi_2$ does not have the problem with resistance change to temperature like SiC, which simplifies the power supply. However, they are very brittle so give them some room to avoid bumping. As with SiC, the elements need clearance from both the refractories and the setting to avoid hot spots. Perhaps the biggest mistake is to buy too small a kiln for general laboratory use. Give the setting and elements some room, and avoid the smallest diameter (3mm) element.

Element life depends on three factors: proper installation, controlled heating and cooling rates, and element bumping. Heating elements are connected in parallel. This increases the amperage of the power supply, but avoids restriction of the element bank by the highest-resistance element.

Thermal Insulation

Fiber insulation has been a boon for laboratory kilns. There is a large decrease in thermal mass and an increase in thermal insulation. Fiber refractories are oblivious to thermal shock. With improved electrical heaters and fiber insulation, the ceramist can determine the firing cycle, instead of the kiln's capability. There are several grades of fibers. Differences are largely a matter of the alumina-to-silica ratio. Higher alumina imparts higher temperature capability. However, it is more difficult to blow high alumina fibers, and they are more expensive. They are available in a variety of forms: block, sheets, blankets, paper, bulk, and rope if one needs it. Lab kilns are usually lined with blocks or sheets. Often, pure alumina rods support the fiber blocks especially for the top. Figure 8.3 illustrates the supporting structures.

Type B is especially useful for roofs with larger spans. Hanger rods fastened to the kiln structure hold a hanger body that in turn supports alumina rods and splines. Unless supported, the fiber blocks will sag at high temperatures. Side panels are also supported in a different configuration C as in Figure 8.3.

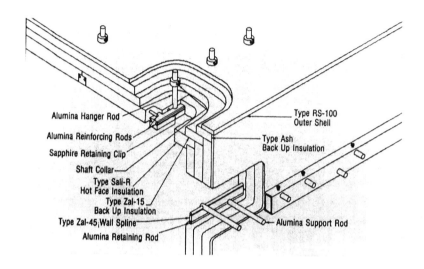

Figure 8.3: Fiber Block Supports. Rigid fiber panels are supported by alumina rods in the structure. (Courtesy of Zircar, TM)

The kiln previously shown in Figure 8.1 has layered insulation as seen in Figure 8.4. Hot face insulation is most critical but, for temperatures above 1600 °C, the fiber block (Zircar SALI (TM)) is currently preferred. Backup insulation consists of an intermediate fiber block with a fiber board against the shell. Note the staggered joints to prevent line-of-sight way for radiation to escape. Again, this is standard practice.

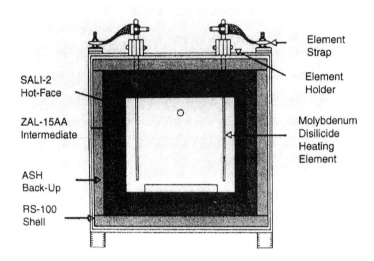

Figure 8.4: Fiber Insulation Structure. Thermal insulation is built up in two or more layers for cost reasons. (Courtesy of Zircar, TM)

Some applications need a hard face when there is a chance for splatters, abrasion, or chemical attack. High alumina refractories are usually a good choice. This type of refractory is not suitable for spanning a distance and the top is an arch, or if one really wants to get fancy, a catenary cross section. If necessary, consult a brick mason as they know the craft.

Traditionally, the backup insulation was fire brick, which comes in several temperature ratings. The higher the temperature rating, the lower its insulating properties. Fiber insulation is becoming almost standard for back up insulation in lab kilns, but not necessarily in production kilns.

Temperature Sensors

The most common temperature sensor is the thermocouple. Most lab kilns use Pt-Pt/Rh thermocouples. At lower temperatures, chromel alumel is very useful because its signal is stronger. It is very important to calibrate the temperature sensors regularly as measurements can drift into error. Thermocouples are calibrated against an NBS standard, from which one creates secondary standards. One can then use the secondary standard to calibrate the working thermocouples. By placing the working thermocouple close to the secondary standard and making a heating run, the two emfs can provide the data for calibration.

W/Re thermocouples tend to be brittle, but, with time, improvements may rectify this shortcoming. The suppliers of thermocouples have instructions for use that specify temperature settings.

One often uses optical pyrometers, especially at higher temperatures where the limit on thermocouples is exceeded. Two-color pyrometers are preferred. The problem with optics is to keep the sight glass clean. This can be done in two ways: wash it with a gas, or use a sight glass assembly that wipes the glass by rotating an "O" ring wiper that cleans it before the measurement.[1] Figure 8.5 shows a sketch of this apparatus.

A gear is built into the rotating mechanism to add a motor to continuously clean the sight glass. It is frustrating to abort a run because the sight glass is fogged up or the sight tube is clogged with deposits. One wastes the entire investment in time and effort when this occurs. To avoid this, put on a self-cleaning sight glass and introduce a gas stream in the tube. This is especially useful for graphite furnaces that when at the desired temperature, exude copious quantities of obscuring materials.

Obviously, the main problem with optics is that they do not function at temperatures below about 700-800 °C. One can use thermocouples for this range and then withdraw them when the optics takes over. Fine tuning of optical pyrometry can get a little complicated when the requirements for temperature control are critical. One can easily avert many of these complications by looking into a black box. All it is, is a refractory block with a hole drilled into it to a depth of about twice the diameter with the temperature reading taken by sighting into the hole. To illustrate this point, read this page illuminated with a halogen light source. The black box eliminates reflections that can mislead an optical pyrometer.

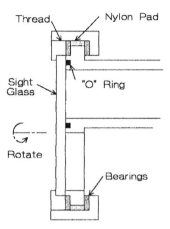

Figure 8.5: Self Cleaning Sight Glass. Rotating the sight glass wipes the "O" ring clean.

Controls

There are many choices of controls. For most uses, the controls are too complicated. One needs a couple of ramps, a soak time at temperature, and a ramp down which often is normal furnace cooling. Whenever the firing cycle is more complicated, a personal computer (PC) is a very good control of choice, but it ordinarily has more capability than needed. It cannot only run the program but can also store the data, analyze the data, signal alarms, store the record, draw graphs, and finally condense a printout. A PC can run a variety of kilns, and they are relatively cheap. A PC is the computer of choice for two reasons: it is less expensive, and there is much more software available. When connected to the appropriate signal conditioners, a PC can do a variety of functions, actuating all kinds of devices.

Hearth

Raise the box kiln hearths to prevent heat from being conducted away from the setting by the hearth plate. A cold hearth will cause uneven sintering. If one has a cold hearth, raise it with an additional refractory plate resting on spacer blocks. These blocks provide an air space under the hearth for hot gas circulation and thermal insulation. The span between the blocks is important, and they should be close enough to resist plate sagging. The hearth plate will still sag a little, but one trick is to put the setting on a bed of coarse refractory grain. If it sags too much, turn it over. Hearth plate materials should be refractory enough, and should be supported enough to resist sagging. Not all alumina are equal, so find one that works. Ask for samples and simply run a sag test by supporting the bar sample over a span and heating to the soak temperature. Measure the sag when cool. Another hearth requirement: it should not react chemically with the setting or the kiln structure. This is highly temperature dependant; Zirconia and alumina are compatible up to 1600 °C. At higher temperatures, alumina will react with stabilized zirconia so one will have to use different setting practices; these will be described in a later section.

Atmospheres

One can use a gasket around the door and a box kiln for firing in other atmospheres, principally nitrogen or argon. Obviously, cool the gasket and securely clamp it to make the seal. When firing in other atmospheres, it is best to evacuate the kiln box and back fill with the gas. Flushing is never complete in a box kiln, especially inside the refractory porosity.

When retrofitting a kiln for atmosphere use, the structure may not be strong enough to support the stresses imposed by the vacuum, in which case the kiln shell will crumple. Remember that gas will expand when heated and pressure will build up. Connect a bubbler to the kiln to avoid this problem. Also remember that the gas will contract as it cools. Never suck water into a hot kiln. A second empty reservoir between the kiln and the bubbler will catch any back flow. One can purchase atmosphere kilns.

Water vapor, CO, CO_2, halogens, and hydrogen are harmful to SiC and $MoSi_2$ heating elements. Firing in hydrogen has the additional problem of being an explosive hazard. Properly designed furnaces are generally safe if maintained. Box kilns are not usually used with hydrogen as they are often not tight enough. An explosion panel is a good idea as it will direct the explosion away from the personnel should the kiln blow up.

Check List, Box Kilns

- Configuration: kiln type and size
- Electrical heating: SiC, Si_3N_4, installation
- Thermal insulation: fiber
- Temperature sensors: Pt-Pt/Rh, optical, clean sight glass
- Controls: simple, PC
- Hearth: chemically compatible, sag, raised
- Atmosphere: vacuum purging, gas flow control, explosion panel

Tube Furnaces, Air

Tube furnaces are usually simple structures with a ceramic tube, a heating element, thermal insulation, and a shell. Most tube furnaces are special purpose furnaces used for a variety of things, some of which will be discussed. It will not be especially helpful to list the furnace types, since these can be obtained from catalogs. Briefly, tubes are made from a variety of materials: dense ceramics, coarse-grained ceramics, fused silica, and metals. Heating elements include: Ni-Cr, Pt, Mo, SiC, $MoSi_2$, and C. The rest of the discussion will focus on problems with air atmosphere, lab-sized tube furnaces, and what to look for when one acquires such a furnace. Figure 8.6 is a simplistic sketch of a typical tube furnace, presented here for reference.

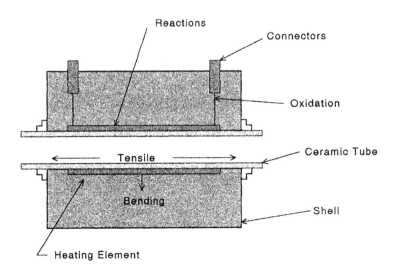

Figure 8.6: Tube Furnace Air Atmosphere. The ceramic tube is subjected to stresses and chemical reactions that can shorten its life.

In the sketch, the tube is subjected to bending stresses if it is not centrally supported. This becomes a little tricky as the conditions at room temperature are different from those at the soak temperature when everything hot expands. Tube failure is usually due to bending stresses at elevated temperatures when the refractory is weak. The worst scenario: when the tube is rigidly supported at the ends and is free to sag in the center. In this case, there are also tensile stresses during cooling, as suggested in the figure, since the kiln structure is cool and does not expand. Figure 8.7 depicts a commercial tube furnace.

Figure 8.7: Tube Furnace, Commercial. Tube furnaces are available in a variety of sizes and temperature capability. (Courtesy of CM Furnaces)

Notice that the tube is free and floating, and though not visible, it is supported on a muffle tube made from a coarser-grained refractory. The shell is split around the tube because this region is hot during a run and has to expand. Otherwise, it would distort and compromise the muffle tube support.

Ceramic Quality

The problems are with sag, thermal shock, strength, and chemical reactions with the ware.

Sag. As the operating temperature increases, the tube will have an increased tendency to sag. There are three ways to cope with this problem. Physically support the tube, especially in the hot zone. Select a ceramic more resistant to sag or from another supplier. Finally, increase the purity of the ceramic tube. Soft glass phases on the grain boundaries result in lower temperature capability. When the glass softens, the ceramic will deform under stress. Central support of the tube reduces the bending stress and helps to stabilize the structure. The manufacturer will have drawings. Ask about the ceramic purity. Most alumina ceramics have some glassy grain boundary phase. These glasses are softer (lower softening temperature) when they contain alkalies than they are when containing alkaline earths. High purity ceramics (99%+) are best, but are more susceptible to thermal shock. SiC refractories have a low tendency to sag. Metals are predictable; there is much more data available on properties with temperature. The Metals Handbook, by ASM, is a good source of data. Nickel alloys are used for air atmosphere tube furnaces.

Thermal Shock. Fine-grained, dense ceramics are susceptible to thermal shock, which can usually be prevented by slow heating and cooling. Thermal expansion and conductivity are important properties. Low expansion and high thermal conductivity result in better thermal shock resistance.

    Modulus of elasticity can be varied by control of the porosity, and in the case of zirconia, by the amount and kind of stabilizer. Ceramic suppliers are a good source of information. Coarse-grained ceramics have a lower modulus and are less apt to fracture by thermal shock. Ask the furnace manufacturer about their ceramic suppliers.

Chemical Reactions. Exposure to certain chemicals can degrade the ceramic tube. In particular, fluxes will lower the softening point. A few such fluxes are Na, K, Li, B, and halogens. One can expect a shorter tube life when these are present. As the temperature is increased, just about everything reacts with just about everything else. One needs to make judicious choices to avoid reactions between the ware, setter, and the tube. The following strategy can be helpful. On a fragment of tube material, place a small fragment of the setter, and on that place a small fragment of the ware. When fired, one can observe these reactions without fluxing down the whole furnace. One can test many combinations in the same run, saving a lot of time. Heating elements are also susceptible to impurities.

SiC and $MoSi_2$ are protected from oxidation by a silica film. One can destroy this film with CO, CO_2, halogens, and hydrogen. An oxide layer on the surface protects metal elements; fluxes can dissolve this oxide layer. The supplier has more information on things that can happen.

Microstructure. Strength of a ceramic part is largely controlled by flaws in the structure. These are introduced during production of the powder and during processing. Some suppliers do a better job of controlling these flaws than others, so the problem becomes one of finding the right supplier. Ask for statistical process control data. If they do not have this data, go elsewhere. One can also request the relationship between strength and grain size. Unbridled grain growth weakens the structure. Heating for long periods at a high temperature can result in grain growth, weakening the tube.

Cements. It is common to fasten the wire heating element to the tube with a refractory cement. Constituents in the cement can attack the element and cause it to fail. Generally, calcium aluminate based cements cause the least trouble with alumina tubes.

Design

There are many things to consider with the kiln design, including things that place stresses on the furnace parts.

Mechanical Restraints. As the furnace heats and cools, the tube must be free to change its size and shape. For example, the tube becomes longer as it is being heated. If it is restrained, it will be in compression. However, this is not the problem. As the furnace cools, it is in tension and it will crack. Ceramics are strong in compression but weak in tension, especially at high temperatures.

Bending stresses. Unless the tube is supported in the hot zone, the sample in the hot zone will place bending stresses on the tube. Ordinarily, this is not a problem as the weight of the sample and its setter is small and the temperature is well below the softening point.

End Seals. All tubes are produced with irregular ends due to break out when diamond sawing. For an air atmosphere, there usually is not the need for an end seal, but if there is, then the tube ends have to be ground flat.

Winding stresses. Often, the heating element is a wire that expands and contracts when heated and cooled. Bear in mind that the wire is constrained by the tube and the cement. To impart slack, wind the coil slightly loose.

Poison Fingers. It has been known for a long time that oils and perspiration from human skin will cause metals to corrode, especially at elevated temperatures. The oils and perspiration from some people are more corrosive than others. Use gloves when handling metal heating elements. If one touches the metal with their bare hands accidentally, clean it with a detergent and rinse.

Leads. Figure 8.6 suggests that the electrical connections are connected to lead throughways that have a substantial reduction in resistivity. It is common practice to double up the wire element from the tube to the lead throughways. All of this helps to confine the heat in the hot zone and to retard the connections from oxidizing. Lead throughways are often refractory metals but can be stainless steel. They are connected by a thread and nut.

Operating Problems

Tube furnaces fired in air are not especially difficult to operate, but they do have some problems.

Thermal Gradients. Since the tube loses heat from both ends, there are bound to be thermal gradients. The uniform hot zone is very short, making it difficult to fire a sample evenly. Often to remedy this problem, the winding is more closely spaced at the ends of the hot zone in an attempt to flatten out the gradient. When the heating element is not a wire, this approach is not available for small furnaces, so keep the sample short or the tube long.

A carriage or boat inserts and holds the sample in the furnace. Typically, these have a short but manageable life. Most problems involve chemical reactions with the sample. An intermediate setter compatible with both is one answer. When there is a liquid phase, use the boat once.

Powder or refractory grain around the sample will adsorb the liquid and prevent a burn through that could destroy the heating element. After sintering, the powder or grain has to be removed. If it sinters hard, this will be very difficult. Coarse-grain packs do not sinter hard so one can easily crumble it off. Some powders, such as BN, do not sinter well and can be removed.

Element Degradation. Metal heating elements will eventually break. Recrystallization of the wire over time will make it brittle. Also, the wire can evaporate, thinning the cross section and making it weaker. Platinum will do this, depositing platinum crystals on the cooler ends of the tube. The main problem is that the element reacts with the tube, cement, or the atmosphere. Metal oxidation is a complex process and is dependent upon the oxygen partial pressure and temperature. For an air furnace element, the atmosphere should be oxidizing.

Check List, Tube Furnaces, Air

- Ceramic quality
- Chemical compatibility
- Stresses
- Refractory cements
- Winding life
- Thermal gradients

Tube Furnaces, Reducing

A good feature of tube furnaces is that one can control the atmosphere by purging. End seals provide a closed volume with an extended length through which one can pass gas. Tube materials include alumina, fused silica, graphite, and refractory metals such as Mo and Ni/Cr alloys. Resistive heating is usually employed, with graphite or refractory metal heating elements. When the tube is impervious, the heating element can be in air and the same materials used in air furnaces can be used for heating elements. Atmosphere composition is often Ar, He, H_2, or a CO/CO_2 ratio. The gas is sometimes chosen for specific chemical reasons dictated by the materials being heat treated. Figure 8.8 depicts a general sketch of a reducing furnace.

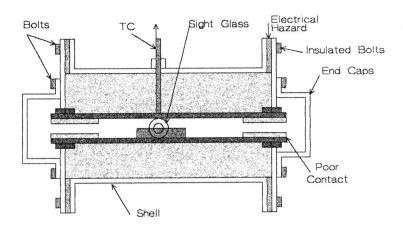

Figure 8.8: Tube Furnace, Reducing Atmosphere. End seals isolate the tube so that a variety of atmospheres can be used to heat treat or sinter samples.

Three subjects will be discussed: materials quality, design, and operating problems. Figure 8.8 will be referred to in the following section.

Materials Quality

Graphite. Furnace parts are made from good-quality graphite. It is useful to ask about the graphite grade in furnace construction and then to give the graphite company a call. Generally, one recommends finely-grained, dense, and fairly strong graphite. Graphite has all kinds of impurities that react with the sample and discolor it. By heating graphite to a high temperature in chlorine, one can remove these impurities. This treatment is commercially available. The entire graphite interior, including the fiber insulation, tube, boat, and refractories can be purified.

Carbon fiber is used for insulation around graphite tubes. This can be bulk fiber or formed parts. Fiber insulation is a lower cost variety, but contains volatile substances that can get into the furnace atmosphere.

Ceramics. One can use impervious alumina or fused silica in neutral or reducing atmospheres. Sometimes, fused silica is called quartz; this is not accurate nomenclature as they are all fused silica. The same criterion applies to reducing furnace tubes and to oxidizing atmosphere furnaces. Make sure that the atmosphere will not reduce the oxide tube.

Metals. A continuing problem with firing in reducing atmospheres is binder burnout. Being in a controlled atmosphere and a closed system, how are the binders removed? Ni/Cr alloy muffles are available and can be installed in a low temperature furnace. The atmosphere in the muffle can be selected and the burnout precisely controlled. The part is very weak after burnout. A good idea is to set the part on the same setter that will be used in the firing. As such, one will not have to handle the part during the transfer to the high temperature furnace. Some furnace systems have gas control mechanisms for handling the burnout reaction products.

Reducing atmosphere furnaces can have refractory metal tubes. These tubes are Ni/Cr alloys for low temperatures or Mo at higher temperatures. Mo is not commonly used in tube furnaces as it is in other types of furnaces.

Stainless steel threads and sliding fits will tend to gall. To prevent galling, one should coat these parts with an anti galling compound. This compound contains copper and aluminum flakes in a greasy material and is available at an auto parts store.

Design

Proper design of the furnace components is very important since manufacturers are not perfect. It is not advisable to assume that the equipment is properly designed just because it was made by a major manufacturer. There is a trend toward SPC and improved product quality, with ISO 9002 certification. This certification is not a guarantee of quality, but it does show that the manufacturer is conscious of quality issues and is complying with some standards.

Purging. The best way to purge a tube furnace is to pump a vacuum and then back fill with the atmosphere. Leaks can be a major problem and may not be easily detected if the vacuum pump is not isolated from the furnace with an isolation valve. With the isolation valve closed, the decay of the vacuum can be observed and, if this happens, there is a leak that can be found and sealed. A cold trap to remove unwanted gaseous species is necessary for attaining a decent vacuum and to protect the pump. It is not necessary to attain a high vacuum for purging and backfill, but the pump should be at least a standard fore pump. With the use of a thermocouple gage and isolation, one can evaluate the vacuum system. It is not uncommon to find leaks in the system, especially on threaded joints. Stainless is hard to thread and a ragged thread has a good chance to leak. Teflon tape is good enough to seal a threaded joint if the thread is well cut. All of the interior furnace materials absorb gases that desorb as the temperature increases. When the atmosphere is critical, the furnace can be pumped down again after heating. This type of equipment is hardly high vacuum technology. The atmosphere is there to control bulk composition and to prevent oxidation of the graphite.

Resistors. For this discussion, the resistor material is graphite. There are essentially two designs: a tube and a slotted cylinder. These are shown in Figure 8.9.

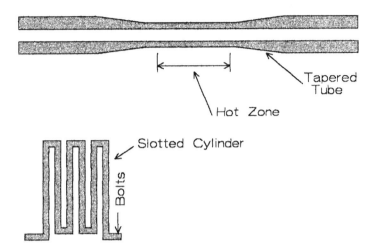

Figure 8.9: Graphite Heating Elements. Heating elements can be in tubular or castellated cylindrical design.

The tube element has a tapered section to increase the resistance in the central part of the tube. This concentrates the heat in the hot zone. It also weakens the tube, but graphite fortunately becomes stronger as the temperature is increased. There are two problems: the ID of the tube is restricted, and electrical contact to the power supply is random. Electrical contact is by a slip joint that may or may not allow the start of the run. When lucky, the contact will occur and as the temperature increases, the tube OD expands to maintain electrical continuity. The ID of the tube is stingy. For a lab furnace, the ID could be as small as 2.5 inches. It takes a lot of amps to heat a low resistance element, resulting in a big installation to fire a very small sample. Slotted hollow cylinders with a flange for connections are shown in the other sketch. In this design, the resistance is greater, reducing amperage. Smaller power supply and positive electrical continuity are substantial advantages. The element is bolted to the power source. Until there is a better connective design, the slotted cylinder is the preferred choice.

End Closures. All slip cast ceramic tubes are diamond sawed on both ends. As the saw blade leaves the cut, the last little bit of the ceramic fractures leaves a small projection. This small projection should be removed by grinding. Figure 8.10 depicts a typical end seal.

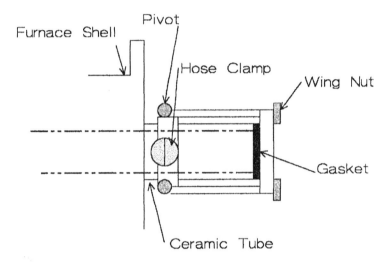

Figure 8.10: Furnace End Closure. A ceramic tube can be sealed with a commercial end closure.

Use a hose clamp to hold onto the ceramic tube. Pivoted wing nuts fit into slots in the stainless end plate and clamp it securely against the end of the tube with a rubber gasket. The little projection that sticks out from the saw cut interferes with obtaining a good seal. The ends of the ceramic tube have to be diamond ground to obtain a seal. Look at the ends of the tube. If there is a sloppy little projection on either end, the ends have to be ground.

In Figure 8.8, the end seals have "O" rings on the end bells, and this works without difficulty. This furnace was built so that it could be pressurized up to 500 psi, principally for firing Si_3N_4 in an N_2 atmosphere. This makes the bells heavy and a little difficult to handle.

Gas Management System. The Figure 8.8 furnace has a poorly designed gas system replete with problems. In the figure, the vacuum system is in the back on the other side of the sight glass. An inlet is for introducing gases through a tube connected to the sight glass assembly. Without an outlet, one cannot sustain the flow. The furnace is pressurized. To sustain a flow of gas, valving is needed to both retain the pressure and to provide a gas flow across the sight glass. This may not be provided, and is something that has to be validated.

Solenoids designed to have the coil activated during the run result in a high amperage in the winding and the coils can run hot, which can quickly destroy them. Compression fittings are a good choice for the gas lines. Pay attention to the manufacturer's directions on tightening. Also, compression fittings are not designed to be dismantled and reassembled. Use a Union fitting where it has to be taken apart.

Solder joints are okay for larger diameter tubing. One needs a little skill to make good solder joints. Make sure that the tube and fitting are not damaged and be sure to clean the parts well. Once the solder flows into the joint, give it a quick wipe with a cloth and leave it alone.

Temperature controls. Figure 8.8 has a two-color optical pyrometer, which is a good choice if the sight glass is kept clean. Optical pyrometers do not work at low temperatures. If the manufacturer did not provide for this, one can make a fitting to insert a thermocouple into the furnace. Seal it with an "O" ring, measure the temperature up to where the optical pyrometer can take over, and then withdraw it. While not automatic, this procedure is tolerable although the temperature measured is on the OD of the heating element and not on the sample. It would be preferable to read the temperature on the sample, but this can be a little hard to do. Programmers can be simple or they can be complex. Operating manuals can be undecipherable, but procedures will work out. Again, a PC is a good choice, especially with the added complexity of gas handling and switching temperature measuring devices.

Operating Problems

Figure 8.8 and another similar carbon tube furnace can have many operating problems. Figure 8.11 is a sketch of the other carbon tube furnace.

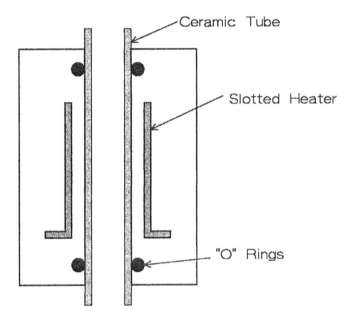

Figure 8.11: Lab Furnace with a Ceramic Tube Accessory. Stresses on the tube during cooling can cause it to fracture.

Graphite is used for all the interior parts. Since graphite oxidizes in air, an alumina tube insert fires in air. "O" rings seal the tube at each end. As the furnace heats, the tube expands along its length placing the hot zone in compression. As the furnace cools, the tube contracts along its length, placing the hot zone in tension. The tube can break and trash the furnace. Friction between the "O" rings and the tube can be excessive. Refer to Figure 8.8 to look at another problem. At the top corner, there is an arrow with the caption "hazard." Voltage across the insulator is low and will not produce an electrical shock. However, if someone is wearing a ring, a metal banded wrist watch, or a metal bracelet, the metal will instantly melt, cutting off whatever appendage it is fastened to. Therefore, avoid wearing any metal object such as a ring or bracelet when operating this type of furnace. The circuit has low impedance, like a welder, so the amperage is high. This is irresponsible engineering. This area has to be at least wrapped with layers of duct tape for insulation, but this is a stop gap measure. Let us hope that anyone else using this equipment will understand the presence of the hazard.

Check List, Tube Furnaces, Reducing

- Materials quality
- ISO 9002 qualified
- Vacuum system
- Resistor design
- Electrical contacts
- Ceramic tube ends
- Gas plumbing and valving
- Sight glass cleaning
- Temperature sensors
- Controller/Programmer/PC
- Tube fracture
- Electrical hazards

Gas Fired Kilns

Ceramic production is predominantly fired in gas-air kilns. Costs are lower than the alternatives, unless they are operating in areas of low electrical rates. High performance ceramics are an exception. In the case where the product is ultimately to be fired in a gas kiln, it can be advantageous to do the lab work in a lab gas kiln. Transfer of a product to production is difficult enough without added complications. Figure 8.12 is of a laboratory gas fired kiln.

This kiln has four burners firing under the hearth. Burners are connected to a blower for the air supply and natural gas to make up the combustion mixture. Proportions of the two are adjustable for either oxidizing or reducing atmospheres. A zirconia oxygen gage measures and controls the oxygen partial pressure in the furnace.

Figure 8.12: Gas Fired Kiln. A wide variety of gas fired kilns are available according to size and temperature capability. (Courtesy of Bickley)

All manufacturers, as far as the author knows, have safety interlocks that turn off the gas supply when there is a malfunction. Flow gages are mounted on the shell for both gas and air. The door is double hinged so the hot face of the kiln will not be in one's face when the door to a hot kiln is open. A Pt-Pt/Rh thermocouple is used for temperature measurement and control. There is a plug in the door to accommodate an optical pyrometer. Automatic controls are available. With this kind of door, the front of the steel shell will overheat as the refractories become worn. This will result in warpage of the shell and a degenerating problem with heat loss around the sides of the door. The door closure refractories have to be maintained. While on vacation, someone failed to maintain the door, and the shell is now warped. What does one do to the kiln? With the door swung out of the way and the refractories covered, slots are cut with an oxyacetylene torch in the shell above the sides and top of the door. Then, the shell is hammered back in place and the refractories are repaired. In the figure, the manufacturer has thoughtfully stiffened the shell around the door with steel gussets. There, warpage has happened before.

Hot gases, called *sting*, exit from the kiln through ceramic tubes located in the top of the kiln. An experienced operator may be able to tell if the burn is going right by the appearance of the sting.

The atmosphere in a gas kiln is a mixture of gases: N_2, H_2O, CO_2, O_2, a little hydrocarbons, and CO. The atmospheric composition can be oxidizing, neutral, or reducing depending upon the ratio of the concentration of these gases. Ceramic parts can be sensitive to this atmosphere.

Several aspects of gas/air kilns include size, configuration, temperature, refractories, controls, and operation.

Size

The size has to be adequate for the parts, but a small kiln will have temperature gradients. It is a good idea to obtain a larger kiln to provide space around the ware and space in the free volume for uniform gas circulation and combustion. For general purposes, three to eight cubic feet is the range for a good lab size. Smaller kilns are apt to have temperature gradients and possibly direct flame impingements on the ware. There has to be space in the combustion chamber for combustion, convection, and heat transfer to occur. Burners should fire under the hearth, and this also takes up space. The top of the combustion chamber is an arch, again requiring space. It is best to not work with small kilns.

Configuration

Dimensions are determined by the shape and the amount of firing. A lab kiln is usually a box with an arched roof. Hearths should always be raised with the burners firing under the hearth. Avoid direct impingement of the flame on the ware. Most lab kilns are front loading, but a bell design has advantages. There are more options in configuration if the firing temperature is not too high. Flatware is often fired on a broad hearth. When the temperature is high, the arch cannot be structurally stable over a broad span. Most lab kilns are shaped internally as a cube with all three dimensions being similar. When there is a requirement for firing long tubes, the kiln has to have a long vertical dimension, which may require additional tiers of burners.

Temperature

Maximum temperature is determined by the burners, combustion gas composition, and the refractories. For the lab, it is a good idea to over design, as the next project may require a higher temperature. Four burners, two on each side, with alternately firing from opposite sides under the hearth are a good design. Fiber blanket refractories are widely used in production kilns, are available in clever attachment schemes, and are usable to about 1600 °C. Lab kilns more commonly use fiber panels. High alumina fiber is available for use up to about 1700 °C. This fiber, available in blanket and formed shapes, is expensive. In the lab, expense is not as controlling a factor as capability. When the hot face is dense 99% alumina brick, the lining is usable to about 1800 °C. For even higher temperatures, the brick is a bonded calcia-stabilized, zirconia bubble.

When properly designed and tuned, a temperature up to 1800 °C is achievable with natural gas/air when large burners are used and the blast gate is open full bore. This is difficult to maintain, but one can either enrich the air with oxygen, preheat the incoming air, or use a gas such as MAP (a high-temperature-burning, welding gas). One can preheat the air by channeling the air through a gap between the hot face and the backup refractories. Design of this structure is critical as support for the hot face can be compromised causing it to sag inward. With sagging, one will need to reline the kiln; this will result in serious delay and considerable expense.

Refractories

The kiln lining has to be engineered so that it can expand and contract. Usual construction is having a wall that holds up an arch through skew blocks with the top free to move up and down. Skew blocks transmit the lateral stresses from the arch through the brick work to the steel frame. Back up insulating brick should be tied to the hot face, and in turn tied to the kiln shell, usually with metal tabs or angled metal supports. These can oxidize and burn off, so it is necessary to make them from a refractory Ni/Cr alloy. Refractories can creep at high temperatures, changing the interior configuration of the wall. The tendency is for the hot face to sag inward, and, if this happens, the insulating brick will be exposed to the hot gases and will fail. Refractories need some maintenance to keep them in place. Refractory suppliers will have data on high temperature properties and hopefully ISO 9002 certification. Low alkali compositions and high density creep less. Burner blocks are castables, generally high alumina with a calcium aluminate bond. Burner blocks have to have a stable location to fire under the hearth without impinging on the hearth supports. Hearth plates are supported with high alumina refractory blocks on a stable base. The plates are a high alumina dense refractory or silicon carbide. High temperature gas kilns usually have high purity (99%+) alumina dense bricks.

Controls

Either thermocouples or optical pyrometers are satisfactory. One advantage of thermocouples is that one can locate them in various parts of the kiln to look for temperature gradients. Optical pyrometers are usually hand held and sight on the setting on the hearth. It helps to have a steady rest to hold the optic on the target. For best results, the optic should sight on a black body, which is a block with a hole having a depth twice the diameter. The burn is programmed with ramps and a soak. A programming problem is that the kiln cannot always keep up with the program, either on the upswing or cool-off. Naturally, the program has to fall inside the limits of the kiln's capability.

Oxygen gages should be standard equipment on gas kilns. The ratio of gas/air is initially set with flow gages. Then, the oxygen sensor can maintain the specified level of oxygen fugacity. It is common to run a little on the oxidizing side when firing oxide ceramics. Pyrometric cones are still useful. They are inexpensive and easy to use. While not useful as controls, they easily reveal the amount of heat treatment adsorbed in various parts of the setting.

Operation

There are several steps to starting up a gas kiln.

Setting. Set the hearth and close the door. (Setting the ware is a sizable subject that will be discussed in a following section.)

Lighting Up. Unlike electric kilns, one starts the gas kiln manually. First, light the pilots. There will be a pilot for each burner, and they will be lit one at a time to establish a slow ramp for binder burnout. One can pull out these pilots to light them before returning them into the burner assembly. Electric spark ignition is also common. Gas kilns have safety features in that the main burners cannot be lit until the pilots are burning. Never bypass safety features; gas/air mixtures can be explosive. Depending on how steep a temperature ramp, the ware and kiln can stand, the burners can be lit one at a time, alternating from side to side and back to front.

Ramp Up. A butterfly valve (blastgate) is set to adjust the amount of air from the blower. The amount of natural gas is determined by the air volume. As the blastgate is opened another increment, the gas volume increases accordingly. The gas-to-air ratio can be fine tuned with information from an oxygen sensor. When the soak is reached, the blastgate is cut back to where the temperature is maintained. Whenever it becomes difficult to hold the soak temperature, the problem might be that the residence time of the combustion gases is too short. By placing a brick partially over the exit port on top, a little back pressure will increase the residence time. For higher temperatures, oxygen enrichment is another option. Some kilns have internal structures where the incoming air is preheated. This will also increase the maximum temperature. Be cautious as some of these structures are not structurally stable and can collapse as the inner lining sags.

Ramp Down. The temperature drop is much faster at high temperatures and it might be necessary to program this with the blastgate down for the initial ramp. At lower temperatures, the thermal conductivity of the insulation usually determines the cool down rate. If everything can stand the thermal shock, one can cool faster by blowing in cold air through the blastgate. A less appropriate option would be to open the door as this introduces cool air and thermal gradients. Whenever this does not damage the kiln or the ware, it is an option. It is a better idea to be patient and let the kiln do its own thing.

Check List, Gas Fired Kilns

- Size: not too small
- Configuration: to suit your needs; bell kilns have advantages.
- Temperature: Make a selection of refractories, burners, oxygen, or preheated air.
- Refractories: Check on quality and structure stability.
- Controls: Select a type. Check safety.
- Operation: binder burnout, ramp and soak, cool down

Other Furnace Types

There are many other furnace types, and it would be just too much to plod through them all. Shuttle kilns, roller hearths, and tunnel kilns are not included, as they are not usually seen in the laboratory. However, one usually fires samples in production kilns of these types. A discussion follows on rotary, induction, and vacuum furnaces.

Rotary Furnaces

These furnaces are used for calcining or sintering granular materials. They are on a slant, with the material fed into the high end and exiting out of the low end as the kiln rotates. The burner is on the low end, firing up into the cylindrical kiln. Electrical heating is more common in lab rotaries. Rotary kilns are dusty and may need a dust collector on the effluent gases. Rotaries are available in lab sizes with metal or refractory tubes. Metals scale with time and will contaminate the batch. Contamination can also come from the seals at the feed end, where metal rubs between the stationary and rotating parts. Another source of metal contamination is from the bull ring and drive that transmit the rotary motion to the kiln. Some designs have chains hanging in the interior for breaking up the material as it sinters. However, this will furnish additional contamination. Just how serious contamination is depends on how much can be tolerated and how much can be removed by magnetic filters.

Material accumulates in the hot zone, making a ring that builds up with time. One uses a shotgun to blast this ring. Large production kilns are even better, as they use a machine gun.

Induction Furnaces

As this is about lab kilns, there are restrictions on size. Figure 8.13 is a sketch of a typical lab induction furnace.

Furnace construction is very simple consisting of the following: a shell, a coil, a base plate, a susceptor, and insulation. The susceptor is often graphite machined out of a billet. The coil is of copper tubing; it keeps the water out of the hot zone in case of a failure. Water is circulated in the coil to keep it cool. Thermal insulation can be graphite fiber, bubble alumina, or bubble zirconia, depending largely on the operating temperature. Tamped lamp black is also used but it is messy. Temperatures above about 1700 °C require graphite thermal insulation. For lab metal melting, the furnace is lined with refractories (often a crucible), and the metal charge can act as its own susceptor. Material for the shell has to be an electrical insulator and should be relatively impervious, especially if the other parts are graphite. Glass fiber-reinforced cement tubing can be used. The coil is connected to a power supply establishing a current in the susceptor and causing it to heat. Optical pyrometry is common for measuring the temperature. One should be careful as voltage and amperage can be deadly.

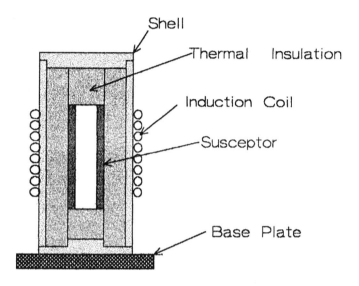

Figure 8.13: Induction Furnace. Lab-sized, induction furnaces are commercially available or can be made in the lab.

Matching the furnace size, coil, and furnace configuration to the power supply takes some experience. Modern induction power supplies are solid state and are easier to match with the coil. Power supplies operate at different frequencies and unless one has the experience, call an expert. Overdesign of the power supply will impart flexibility that gives one a wider choice on furnace design and size. Suppliers of the power supply can help with these choices. Many lab induction furnaces are home made, probably because they are so simple. There are many other uses for induction furnaces, mostly in industry. This discussion is abbreviated to the most common experiences in the lab.

Vacuum Furnaces

Some materials have to be fired in a vacuum or in a protective atmosphere. Equipment for this is commercially available and is complex, versatile, and expensive. However, it is better buying this equipment from a supplier because of the extensive engineering and experience involved. Cosider the following summary.

Size/Configuration. Hot zones are one to three cubic feet. Shells are stainless steel and are cylindrical with hemispherical end bells. The door swings out, giving front access. The door can be vertical or horizontal, but it is usually horizontal.

Temperature/Cooling.   Ni/Cr 1000 °C
Mo 1650 °C
C 2200-2600 °C
W 2500 °C

Insulation/Shell Cooling. Graphite fiber, Mo multiple heat reflectors. The shell is cooled with a water jacket, which can be recirculating. Forced gas cooling is available through a heat exchanger.

Vacuum/Pressure. Mechanical fore pumps, diffusion, turbo, or cryogenic high vacuum pumps/Pressure to 2500 psi

Atmosphere. Vacuum to 10^{-6}Torr (.013Pa), H_2, N_2, Ar, and air with Ni/Cr.

Burnout. Binders and wax can be burned out with special equipment.

Controls. Modern computer controls, interlocks, and alarms

Research.   Obtain a customer list and personally check out the performance of identical or very similar equipment.

Check List, Other furnace types

- Rotary: temperature, size, tube material, clean out, contamination
- Induction: Get expert advice on size, materials, power supply, and safety.
- Vacuum: size, temperature, materials, heating elements, atmosphere.

3.0 SETTING PRACTICES

Formed ware is set on the hearth in many ways, depending on composition, sag resistance, shrinkage, size, and shape. The best case is as follows: the composition of the hearth and ware is identical, the part does not sag at the soak temperature, the part has no or very little firing shrinkage, and the part is small with a blocklike shape. Such hypothetical scenarios do not reflect real life, however. This section describes ways to set the ware on the hearth of the kiln. Often, conditions are less than ideal. For example, the hearth reacts with the ware, both the ware and hearth sag, the ware shrinks up to a fraction of its green size, or the ware is huge, fragile and has a thin cross section. Firing in air is often different from firing in an atmosphere and will be discussed separately.

Firing in Air

Setting practices follow the same principles for a variety of kiln types, except that the size and configuration can change.

Composition

Isolate the body when it reacts chemically with the hearth. These reactions are often predictable, from the free energies of formation. When venturing into new materials, it is worth the time to look up this data. A test run with a small sample and setter is a good idea. A bed of chemically compatible grain spread on the hearth is a common way to isolate the part. Cover this bed with fresh grain for each run and level it. Use relatively coarse grain as it will not sinter hard or shrink. Wiggle the parts into the bed to seat them. When there is an ID, do not get any grain into the center as it will prevent shrinkage and will warp the part. After the thickness of the bed builds up, it is scrapped off and the process is repeated. One can also protect the hearth with disposable plates partially compatible with the chemistry. The area of direct contact between the plate and hearth is small, limiting the extent of interaction. One can also set parts directly on compatible plates. This is common when firing granular materials where a box or sagger is lined with plates, isolating the ceramic from the hearth.

Test runs will help to choose compatible materials. Alumina and zirconia are compatible at 1500 °C but not at 1700 °C. Some newer advanced ceramic powders sinter at lower temperatures, making setting composition simpler. Another related way is to bury the parts in a bed of grain. $BaTiO_3$ capacitors are fired this way. The composition of the ceramic relates directly to its properties: thermal conductivity, thermal expansion, elastic modulus, and strength (if we include microstructure). The worst case is a ceramic with low thermal conductivity, high thermal expansion, a high modulus, and low strength. This material is hard to set properly because it is highly susceptible to firing conditions. High internal stresses and thermal gradients induced by uneven heating can cause it to fail by distortion or fracture. For example, a fine grained ZrO_2 part will be a lot more difficult than an identical shape made from coarse grained MgO. Setting to even out temperature gradients helps to alleviate this problem.

Sag

When the composition contains a glassy phase, it is more likely to sag. Also, fine-grained materials are more likely to sag and distort than coarse materials. The glass composition in the body will determine its softening point and viscosity. Less glass, a higher softening point, and higher viscosity will reduce sag. If there is this option, the composition can be adjusted. If not, one can support the part; in a grain bed, set in a V-shaped block, stacked with a little grain between layers, and confined physically by refractories often coated with a wash. A wash is a slurry painted onto the refractory surface that is then let to dry. The composition of the wash should be compatible with the ceramics being fired. Washes are not too successful as they bleed through, but one can still use them. If this is admissible, one may reduce the temperature. While not a usual laboratory procedure, the part can be set on a specially formed setter for firing. Setters are common in industry but not so much in the lab because the parts vary so much in size, composition, and shape. Another instance of sag is where the part is formed as a fluid, such as with injection or wax molding. When the part is reheated to fire it, the polymer or wax remelts and the part can slump. A common remedy for this is to pack the part in a powdered absorbent and sagger for debindering. As the binder melts, it is absorbed in the pack, leaving the green ceramic that can then be removed and fired. If the polymer is crosslinked, one can fire the part directly as it will not remelt. Binder burnout has to be very slow or the part will burst from internal gas pressure. Guidance can be provided by a thermal gravimetric analysis (TGA) on the binder. Over the region where the binder is losing weight rapidly, fire the part on a slow ramp.

Shrinkage

Consolidation of the body can result in shrinkage. Shrinkage is a problem because the part is moving, introducing distortions and stresses. Coarse-grained materials are usually packed to where the coarse part of the composition is in grain-to-grain contact, and will not shrink appreciably. Shrinkage and bonding occur internally in the fine-grained, bonding portion of the body. One can stack these low-shrinkage materials in the setting. Stacking along the edges helps inhibit cracking by keeping the material in compression. Sometimes, it is laid flat, which limits the height of the stack. It is necessary to provide gas circulation throughout the stack. Groups of parts are set crosswise to provide open spaces in the setting. Ceramics that have appreciable shrinkage are more difficult to set on the hearth. Fine-grained ceramics commonly have up to 20% linear shrinkage. One usually uses short stacks of three. The part is moving on the hearth as it shrinks, resulting in friction on the base. This friction introduces a counter force in the opposite direction from that due to sintering. As a result, the base of the fired part is wider than the top. One often uses shrinkage plates to address this problem. Figure 8.14 illustrates a typical use of a shrinkage plate.

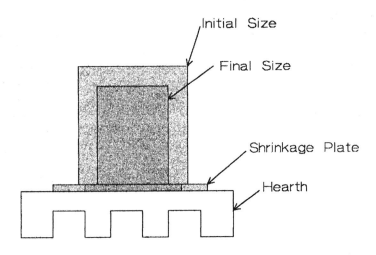

Figure 8.14: Shrinkage Plate Setting. Use of a shrinkage plate relieves basal shrinkage stresses on the part being fired.

Since the part and plate shrink together, the drag is at the hearthplate interface and not at the part-plate interface. Shrinkage plates are expendable, with a new one used every run. They are usually the same composition as the part. The cost of a plate is not a big factor in the lab, but it can be prohibitive in production. Even then, there is no other reasonable choice. Figure 8.15 illustrates how a part distorts without a shrinkage plate.

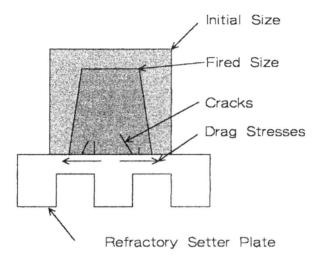

Figure 8.15: Sintering without Shrinkage Plate. When a shrinkage plate is not used the part can distort and crack.

There are two opposing forces on the interface: the sintering force pulling toward the center and the drag force, due to friction, pulling in the opposite direction. Sintering is inhibited on the base of the part, so it shrinks less and has a lower density. It can also crack when the drag exceeds the strength of the material. This sort of problem is exacerbated when the part sticks to the hearth, often due to a glassy phase weeping out onto the hearth. In the last two figures, it was assumed that the green part had a uniform density throughout. This is a false assumption when the part is die pressed. In this case, the pressure in the center is less, the green density is less, shrinkage is greater, and the part hourglasses. This effect is shown in Figure 8.16.

Hourglassing is always present for die pressed parts that have appreciable firing shrinkage, unless they are plates. Otherwise, one can diamond grind the fired piece or change the forming method.

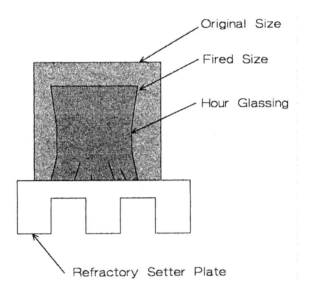

Figure 8.16: Hourglass Distortion. Pressure gradients, during pressing, increase the firing shrinkage in the middle of the part causing hourglassing.

Size

Large pieces are more difficult because of their mass and internal temperature gradients during firing. Coarse materials are much less limited, as they have little shrinkage. Fine-grained materials that shrink are size limited, and anything larger than about a 4-inch cube or an 8-inch-diameter plate is tough to make. Best chances are to set the piece on a shrinkage plate and fire slowly. A monthlong firing is not uncommon. Set the part so that it heats evenly. Make sure that the heat can surround the piece, and it is common to enclose the part in a covered sagger, sometimes buried with coarse grain. The idea is to minimize thermal gradients in the part. When a gradient occurs, the outside will start to shrink sooner than the interior, setting up stresses. A big piece where thermal diffusion takes longer gives the most problems. Most cracks occur during forming, drying, or heating, often during binder burnout. Incipient cracks will open during firing despite the setting and ramp schedule. Looking at the crack edges with a microscope will provide information about when the crack occurred. When the edges are rounded, the crack was present before the soak. If the edges are sharp, the crack occurred during cooling.

Shape

When the shape is complex and fragile, the setting will support the piece during firing, which is often done in a bed of grain. Keep in mind that as the part shrinks, the grain pack will not. Grain packs do not flow well. Spherical ceramic beads can be used as a pack as they will flow a little better than angular grain. Additionally, one can use sintered, screened, coarse, and spray-dried spheres as a pack or setting sand. One can also use a small, 2-mm-diameter, spherical, grinding media. When grain is trapped in an interstice, part shrinkage will be restrained and the part will distort or break. It is essential for the part to be free to shrink. It is difficult when the shape has both thin and thick cross sections, such as a turbine rotor. Figure 8.17 illustrates an idealized, tough shape.

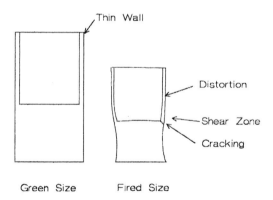

Figure 8.17: Thin/Thick Wall Part. It is difficult to sinter a part with radically different cross sections due to warpage and cracking.

Thin sections of the part will start to sinter sooner than thick sections, resulting in distortions and cracking. The thick section will shrink after the thin portion is all done, putting stress on the shear zone..

Another tough shape is a thin flat plate, shown as it sinters in Figure 8.18.

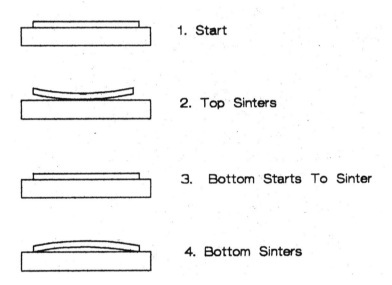

Figure 8.18: Sintering Thin Flat Plate. A thin plate tends to warp during sintering due to uneven heating.

A ceramist was trying to produce a flat capacitor part and had tight specifications on flatness. The plate would come out of the kiln sometimes curved up, and sometimes curved down. A perplexing problem? After trying a number of options, it was decided to observe the part during the firing process as in Figure 8.18. First the part lays flat as in stage one. Then it curved up as in stage two. It flattened out as in stage three. Finally, it curved down for the final shape as in stage four. The best solutions are often the simple ones; the part was taken out of the kiln at stage three. The top of the part was heating faster than the bottom, causing the top to shrink and bending it upward. As the bottom starts to sinter, it shrinks and causes the part to flatten out. As the sintering continues, the part then curves the other way. Sometimes, timing is everything. Apparently, there was a cold hearth or a hearth plate with too much thermal mass contributing to the problem, but the solution was simply to remove the part at the time when it was flat.

Thin parts present special problems. A thin plate of tetragonal zirconia polycrystal (TZP) that is yttria-stabilized zirconia can be die pressed and fired. The problem is that the part is thin, only 0.040 inches in thickness and 6 inches in length, and it is bound to warp. Thin pieces of TZP can be set between two 99% alumina setter plates that are one half inch in thickness and are diamond ground flat and smooth. Carefully place the pressed green strip between the two plates and fire it. The TZP part comes out of the kiln flat and intact. However, it did warp laterally probably due to an uneven die fill.

Setting Long Tubes. A short, stubby tube can be set on end, but a long tube cannot. It could be set in a V-shaped block, but there is a good chance that it will sag out of round. A better way is to hang fire the tube as in Figure 8.19.

In the kiln, a support structure at the top will hold the tube as it sinters. At first, the tube is held up by a block at the bottom. As the binder burns out, the tube gets very weak and cannot support its own weight. However, since it rests on the block, it is in compression where it is much stronger. It gets stronger as it starts to sinter. Shrinkage lowers the tube onto its conical seat, and it separates from the base support. Now, it is in tension; however, it is strong enough to withstand the stress. Being in tension, the tube straightens as it fires and retains its circular cross section. Castable alumina refractory can be used for the support structure. The tube shown in the figure has a conical top that is slip cast as an integral part of the tube. Another way is to cement a collar on the top of the tube to hold it during sintering. For thermal uniformity up the length of the furnace, there might have to be more than one tier of burners. For hang firing, the burners fire in tangentially to avoid direct flame impingement.

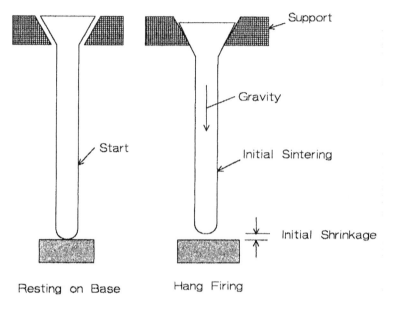

Figure 8.19: Hang Firing Tubes. Hang firing keeps tubes straight and round.

Check List, Setting in Air

- Chemical compatibility
- Setting bed
- Setting plates
- Setting pack
- Supports
- Binder burnout bed
- Stacking ware
- Shrinkage plates
- Setting large parts
- Setting special shapes
- Hang firing long tubes

Setting for Atmosphere Firing

Two aspects of setting atmosphere firing will be considered: graphite furnaces and metal furnaces.

Graphite Furnaces

Two ceramic materials are predominantly fired in graphite furnaces: SiC and Si_3N_4. A problem with both these ceramics is that one constituent has a much higher vapor pressure than the other. Silicon has a much higher vapor pressure than carbon, and nitrogen has a much higher vapor pressure than silicon. These ceramics will partially decompose during sintering. Setting practices can compensate for this problem in a few ways.

Enclosures. One can enclose the part in a graphite box to increase the local vapor concentration of the more volatile species. This works for SiC, especially if the box has been pre-siliconized. It also helps if one buries the part in SiC grain. One has to use coarse grain (60 mesh or so) or else the grain will sinter hard making it too difficult to remove the part later. Figure 8.20 depicts a typical enclosure.

This box is machined from a block of graphite with an interlocking lid. Ordinary graphite has impurities that can be removed by heating in Cl_2, a commercially available process. One can pre-siliconize a furnace or container by first firing a bed of SiC grain, which coats the interior with SiC. It is not a good idea to use Si for this process as it will cut a hole through the bottom of the box when it melts. Silicon can be added to SiC grain as there is enough grain to contain the molten Si by capillary forces.

Figure 8.21 shows another type of enclosure.

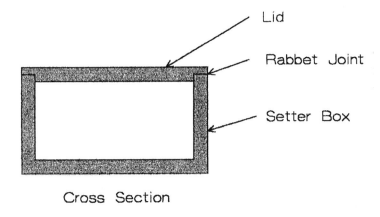

Cross Section

Figure 8.20: Graphite box Containment. Samples are often enclosed in a box to control the sintering atmosphere.

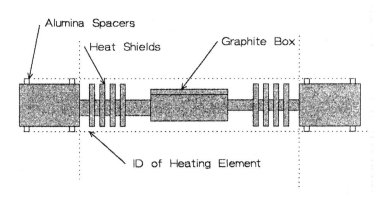

Figure 8.21: Tube Furnace Support. A firing assembly can be inserted into a tube furnace to hold the parts being sintered.

The two end members of graphite have rods of alumina that electrically isolate the support from the heating tube. Threaded rods hold the boat in the hot zone. The boat can have a lid as Figure 8.20. While the structure seems fragile, graphite is quite strong and creep resistant at high temperatures. For a vertical tube, one can use a structure such as that in Figure 8.22.

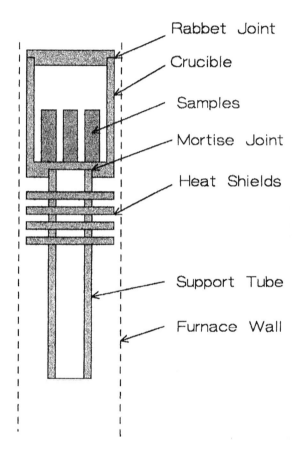

Figure 8.22: Vertical Support. With a vertical tube furnace, a sample holder can be inserted into the hot zone.

Fins on the support rod radiate heat back into the hot zone and help to keep the end cool. A graphite crucible holds the sample.

Graphite plates

When the furnace is larger, one can set the part directly on a graphite hearth plate, as in Figure 8.23.

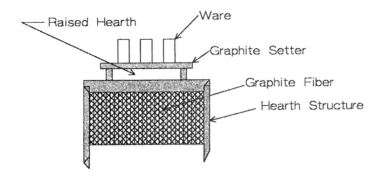

Figure 8.23: Graphite Hearth. The hearth should be raised and on a base insulated with graphite fiber.

One can also place a box on the hearth plate. Avoid cold hearths. In Figure 8.23, the plate is raised to allow for even heat distribution. Graphite fiber is an excellent insulating material; one can place it in the hearth made from thin graphite plates.

When firing SiC exclusively, it is necessary to siliconize the entire furnace in a prefiring. This can be done by firing SiC grain or SiC grain with Si mixed in. A graphitic layer forms on the SiC part after firing, if it is not set in a Si-rich atmosphere.

Si_3N_4 is fired in a N_2 atmosphere, often at a high pressure to suppress nitrogen loss. About 200-1000 psi is in a usable range depending upon the temperature. Nitrogen loss is a more serious problem as the temperature increases, as this requires a higher pressure. One can do several things with the setting.

Pack the part in S_3N_4 grain, in a graphite box lined with BN plates, or in a box coated with a BN wash. BN is hot pressed and is very expensive.

Metal Furnaces

There are two classes of metal furnaces: Ni/Cr alloys are for low temperatures, and Mo is predominantly for high temperatures. With Ni/Cr, one can buy muffles that slip into preexisting furnaces and can be connected to vacuum/atmosphere manifolds. Binder burnout of materials, where this is inadvisable in the sintering furnace, can be done in the muffle.

One can use Tungsten as heating elements whenever very high temperatures are necessary because it embrittles. Tantalum (Ta) also embrittles but it is very expensive.

Thin sheets of Mo or Ta can be used as setters, often just once. One can also use refractory ceramic plates such as alumina, magnesia, or zirconia as setters, but make a nondestructive trial run as a precaution. As mentioned earlier, free energy data can be used to predict reactions between materials. If one is not familiar with this, an expert should be consulted.

Check List, Firing in Atmosphere

- Volatilization
- Purified graphite
- Support structures
- Pressurized atmosphere
- Metals, Ni/Cr, Mo
- Metal foil setters
- Setting on graphite

4.0 FIRING PROCEDURES

After proper milling, slip preparation, and part formation to a uniform high density, the difficult tasks will have been done. With good equipment in place and the parts set properly, firing is usually straightforward. There are four steps in firing for most ceramics: binder burnout, ramp up, soak, and ramp down. Each step will be considered, as it was before for firing in air.

Binder Burnout

There are two potential problems, bursting the part and, with injection or wax molding, the part may slump. With bursting, the problem is having permeability high enough to allow the binder decomposition products to escape without building up excessive internal pressure. Obviously, this is less critical with coarse materials than with fine. Ceramics with low thermal conductivity are also more difficult because of thermal gradients. Modern programmers make burnout easy, and ramps of about one half degree per minute up to 500 °C are typical. If there is trouble, just adjust the heating rate. Most binders are essentially gone by 300 °C to 400 °C, which can be determined by thermal gravimetric analysis.

Ramp Up

With the binder burnout, the heating rate can now be steeper. Ramps of 4-5 °C/hr are typical for fine-grained, dense ceramics where thermal shock is a problem. Fire the other materials on a faster ramp. During ramp up, porosity is coarsening and sintering starts. High rates of grain boundary movement can result in pores being trapped within the grains, which suggests that the upper part of the ramp might be too fast for fine-grained, dense ceramics. Atmosphere in the pores can also have an effect. Gases that can diffuse along grain boundaries help to attain a high density. These include oxygen, hydrogen, and vacuum for ceramics such as alumina. Gases such as nitrogen and argon are more difficult to remove.

Soak

Determine the soak temperature and time experimentally. This is another good place for a statistically designed experiment, where the independent variables are milling time, die pressure, soak temperature, and time. The dependant variables are chosen to optimize the properties of the ceramic. These can be density, strength, permeability, or whatever is needed. Lab firing cycles generally have a soak time of about one to two hours. Pure oxides soak between 1400 and 1800 °C, depending on the particle size and composition.

Ramp Down

The most common ramp down is just to turn the power off and let the furnace cool at its own rate. The first part of the cool down is steep, and it might be necessary to fire down for the initial part of the ramp. Generally, there is little difficulty associated with ramp down, except the newer fiber lined kilns where the ramp could be too steep, causing thermal shock.

Multiphased Ceramics

These materials can present special firing problems. Two systems will be considered as challenging: MgO stabilized zirconia, and alumina with an interstitial glassy phase.

Partially Stabilized Zirconia

Zirconia by itself undergoes a phase change at 1200 °C, from monoclinic at lower temperatures to tetragonal at higher temperatures.

There is a big increase in density, which disrupts the ceramic to where it falls apart. To avoid this, the zirconia is stabilized. CaO, Y_2O_3, CeO_2, or MgO are the most common stabilizers. There are other less important molecules such as the rare earths. The ZrO_2-MgO system is especially interesting. It is a commercial material that shows the use of ceramic crafts. The phase diagram is as in Figure 8.24.

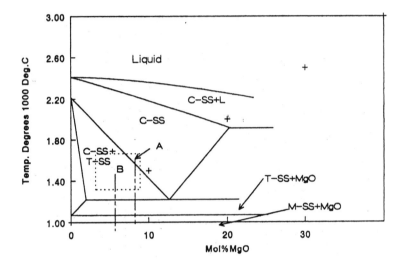

Figure 8.24: ZrO_2-MgO Phase Diagram. The part sintered at A is in the cubic solid solution region. When cooled, it will precipitate out small inclusions of tetragonal zirconia and impart toughness. The part sintered at B will contain both tetragonal and cubic phases. This imparts a lower thermal expansion and improves thermal shock.

Along the horizontal axis is the amount of magnesium oxide added to the batch. When this batch is heated to an elevated temperature, several particular phases form. The diagram shows these phases: C-SS cubic solid solution, T-SS tetragonal solid solution, and M-SS monoclinic solid solution.

Solid solutions are where the material crystallizes in a specific crystal system, but the amount of MgO is variable within the limits shown in the diagram. The region where tetragonal and cubic coexist is of particular interest and commercial importance, and will be discussed at some length.

Fully stabilized zirconia with MgO is a brittle ceramic of little commercial interest. However, if it is only partially stabilized, the ceramic system is of great interest. The area of concern in Figure 8.24 is the dashed-lined box containing two compositions marked by vertical dashed lines. In the lower part of the diagram, are the T-MgO and M-MgO regions. Decomposition of the solid solutions is sluggish and does not occur in a normal cool down. One does not generally find MgO. Long aging in these regions will cause decomposition, but this does not normally occur in conventional processing.

Look at the composition on the right. Zirconia is batched with magnesia, milled, dried, and pressed into a part such as a small disc. It is set and fired above the phase boundary into the cubic solid solution region. This temperature is dependant on %MgO, becoming higher as the MgO content decreases, as shown in the phase diagram with the cubic boundary sloping up to the left. In the sketch, the composition was selected to make this temperature about 1800 °C, which is about as high as a gas kiln can fire unless one has special equipment. Even then, oxygen enrichment may be needed in a gas-fired kiln. The sample will sinter to near full density if prepared correctly. It will be monophased cubic with a coarse (40-60 μm) grain structure because of the high temperature. The phase diagram, and not the particle size of the starting powder, dictates the temperature. Unlike other ceramics, the size of the powder particles influences the sintering temperature. However in this example, the temperature is a consequence of the composition not the particle size.

As the ceramic cools, it drops into the C-SS, T-SS area, and the tetragonal phase precipitates out as small (about 0.2 μm), lenticular particles enclosed in the cubic matrix. One can predetermine the tetragonal phase amount from the phase diagram. As expected, there is less tetragonal when close to the cubic boundary, and more when further to the left. This can be estimated by the inverse lever law. Draw a vertical line at the body's composition in the phase diagram. Next, draw a horizontal line at the selected temperature and measure its length in the C-SS, T-SS phase region to the vertical composition and measure the other side up to the left boundary of the phase region. The inverse ratio of these lengths is the equilibrium ratio of the amounts of the two phases. Using transmission electron microscopy, Heuer photographed this structure.[2] This is shown in Figure 8.25.

Figure 8.25:   Tetragonal Zirconia Inclusions. Tetragonal zirconia inclusions in the cubic solid solution phase impart toughness to the ceramic. Scale bar 0.1 μm.

Look again at the phase diagram. As the temperature decreases, the amount of the tetragonal increases. To reach equilibrium, the ceramic has to be aged at an intermediate temperature such as 1500 °C for about an hour. One can craft the ceramic using the tools in the following list:
- original composition,
- sintering temperature,
- sintering time,

- annealing temperature, and
- annealing time.

Use these tools to craft the ceramic and to optimize the properties that one seeks. After identifying the independent variables, one can craft a statistically-designed experiment and optimize the ceramic manufacturing process.

Crafts are not only related to physical work, but also to a mental exercise. Mental work is central to any craft, especially ceramics. For example, consider an experiment on sintering where the high strength is the objective. Selection of the following independent variables, die pressure, % MgO between 10 and 16 mol % with ZrO_2, and a sintering temperature between 1500 and 1700 °C, obviously positions one at the wrong target. After looking at the phase diagram (Figure 8.24) and reading the literature, one would realize that the material is brittle and in the cubic phase region.

Fracture toughness is an important property. If one uses the tools effectively, the ceramic can be toughened. Tetragonal zirconia inclusions within the cubic crystal lattice want to transform to monoclinic, but they are restrained from doing so. As a crack passes through the ceramic, a halo of stress relief precedes the crack tip; the transformation takes place within this halo causing an expansion in volume and squeezing the crack tip shut. Since the crack can no longer proceed, the ceramic is toughened. This is called transformation toughened zirconia. A small sphere of this material can be hammered into a steel block without it breaking. There is extensive literature on transformation toughened zirconia, but extensive detail is beyond the scope of this discussion.

Now consider the composition on the left. Figure 8.24 is repeated below as Figure 8.26. The MgO concentration is lower as it is further to the left. Also, the ceramic is sintered in the two-phase region of the diagram. There will be two phases present as separate grains: C-SS and T-SS. As the ceramic cools, the C-SS starts to decompose as before, precipitating tetragonal inclusions within the cubic grains by rule of the inverse lever law. Tetragonal zirconia occurs in two habits: free standing grains and then, as before, tiny inclusions within the cubic phase. These will transform into monoclinic zirconia at two different temperatures, doing wild things to the thermal expansion curves. Figure 8.27 is of this thermal expansion curve.

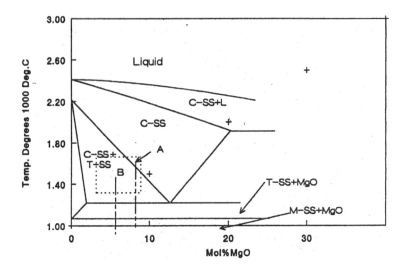

Figure 8.26: ZrO₂/MgO Phase Diagram. (Figure 8.24 is repeated)

While Figure 8.27 is dramatic in appearance, the shape is a blessing. When compared to fully-stabilized zirconia, this material has a radically reduced thermal expansion. Thermal shock is excellent. One can heat this fine-grained body to a red heat and plunge it repeatedly into cold water without breaking it; however, it will eventually, after 20 to 40 cycles, fatigue and weaken. Tubes of this material have been used as thermocouple protection tubes and can withstand the shock from being plunged directly into molten steel. Additionally, one uses such coarse-grained refractories in foundries where one pours molten alloy directly across the cold lip of the crucible following a melting process by induction heating.

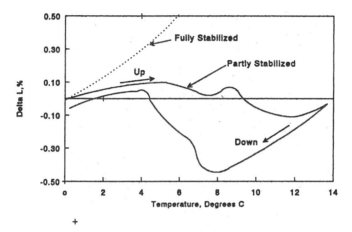

Figure 8.27: ZrO$_2$-MgO, Thermal Expansion. Referring to Figure 8.26, the thermal expansion of B containing both free standing and included tetragonal-monoclinic zirconia produces a thermal expansion with a low value.

Why is there such an odd shape to the curve? The first hump in the curve is due to the freestanding, monoclinic grains transforming to the tetragonal phase. This results in a contraction. The second hump or contraction in the curve to the right denotes the internal, monoclinic inclusions transforming into tetragonal zirconia. Tetragonal grains enclosed in the cubic matrix temporarily restrain them from flipping over to the tetragonal crystal structure. This restraint on increasing the inclusion's volume is responsible for the higher transformation temperature. Overall, the body has a low thermal expansion, which is great for thermal shock resistance. In referring to the list of craft tools, one recalls the sintering temperature. Note, on the phase diagram, that the sintering temperature is lower than that for the first composition. Typically, it is about 1650 °C. At this temperature, zirconia will not fully densify leaving residual porosity. When fully dense, the ceramic produces a high-pitched ring if tapped with a steel rod. When porous and with a lower modulus, the ceramic produces a flat sound. This is exactly what one needs. The flat sound means that the modulus of elasticity is lower, greatly increasing resistance to thermal shock. The two tools, much lower thermal expansion and lower modulus, enable the ceramic to sustain serious thermal shock for about 20-40 cycles.

Take another look at the phase diagram. It does not take very much MgO to partially stabilize the structure. This makes the ceramic sensitive to chemical changes in composition. For example, when silica is present, it reacts with the MgO, shifting the zirconia composition to the left. Use X-ray diffraction measurements of the ratio of cubic to monoclinic (or tetragonal) to establish a location on the phase diagram. In the example, one can add MgO to correct the composition.

Check List, Zirconia

- Study the phase diagram.
- Select the amount of MgO.
- Establish the firing curve:
  Binder burnout,
  Ramp up,
  Soak,
  Anneal, and
  Ramp down.
- Correct the composition.
- Correct the firing cycle.

Al_2O_3-glassy bond

These ceramics are commonly used as wear-resistant materials, grinding media, electronic substrates, and jar mills. The microstructure is unique with small (2-4 μm) alumina grains faceted and embedded in a glassy matrix.

Composition. Alumina is present as 85-90% of the composition. Glass is formed by batching in clay, talc, and sometimes $CaCO_3$ or $BaCO_3$. Clays are usually ball clays. The glass formed during firing is an alkaline earth alumina silicate with Na_2O picked up from the Bayer Process alumina.

Sintering. Liquid in mass transport also helps densification. This allows a lower temperature (1500-1550 °C or so) for sintering; 1500 °C is cheaper than 1700 °C that may be needed for pure Bayer alumina. Burnout, ramp up, soak, and ramp down are not unusual.

Microstructure. Some regions in the structure are fully dense containing faceted alpha alumina crystals in a small amount of glass matrix. In other regions, there are large pores 10-20 μm in diameter. It is hard to prove, but it is possible that these pores form when the glass coalesces to form an impermeable mass. Then, constituents in the composition continue to degas, resulting in the bubbles that we call porosity. If these speculations are close to being right, then possible solutions could be to soak at a temperature just below that where the gases come off, to substitute non gassing raw materials, or to fire in a vacuum. Dental porcelains are vacuum fired to remove porosity. Vacuum firing is not practical for bulk amounts of a product. Substitution of raw materials could be a more practical solution. Perhaps, it is the ball clay causing the porosity. Ball clay contains all sorts of gas precursors such as water and organic particles. Kaolin eliminates the organic component and calcined kaolin eliminates the water. A few experiments on a degassing soak might be worth considering.

Checklist, Firing alumina/ Glass

- Temperature
- Gas evolution

Recrystallized SiC

SiC is a very tightly, atomically-bonded material that does not shrink when sintered unless additives are included. Whenever the sintering mechanism is either vapor transport or surface diffusion, the ceramic recrystallizes but does not shrink. Strength is higher with higher densities. As it does not densify, it is necessary to obtain a high green density during forming. This is achieved by gap sizing and slip casting.[3] Figure 4.29, shown earlier, indicates the general area for obtaining the maximum green density. Casting slips have a high solids content and are dilatant. Deflocculant concentration and pH are both controlled.

As previously mentioned, the kiln is set to a Si-rich atmosphere realized by either siliconizing the furnace, setting in SiC grain, or both. The kiln response determines the shape of the firing curve. A typical case is a curved ramp up to 2200 °C in three hours, a 2-hour soak, and a cooling curve down to 800 °C in less than two hours. The furnace can include a large power supply for a rapid ramp up and a gas circulating system with a heat exchanger for a rapid ramp down.[4] Use a vacuum to purge the furnace. Two vacuum stages make the pump down more efficient: a blower and a fore pump. The blower in front of the fore pump makes the fore pump more effective.

Check List, Recrystallized SiC

- High solids slip
- High green density
- Siliconize furnace
- Purging

5.0 HOT PRESSING

When forming materials, there are three principal, controlling factors: temperature, time, and pressure. Figure 8.28 places some perspective on these parameters.

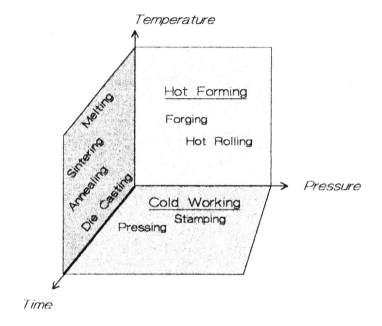

Figure 8.28: Forming Methods. On a temperature/pressure/time diagram various forming methods are plotted.

Typical processes are shown in the Figure 8.28. The pressure-temperature plane locates the hot forging and hot rolling processes. Time is a minor factor in this plane. The time/temperature plane locates melting, sintering, die casting, and annealing processes. The pressure/time plane is where the cold working processes appear, such as pressing and stamping. The volume of space is the domain of hot pressing. For ceramics, typical ranges of temperature are from 1300-2700 °C, depending on the ceramic and particle size. Pressures are between 2000-10000 psi and are not strictly limited to these values. Alumina can be hot pressed from 1400-1700 °C and pressures from 2000-6000 psi in usual cases. With the simultaneous application of both temperature and pressure, the ceramic particles sinter at a higher rate producing a fully dense body at a temperature lower than that used for sintering alone. A finer-grained microstructure is obtained, usually resulting with higher strength and density. While true for most materials, the very high quality, submicron, ceramic powders now available can sinter to structures rivaling those produced by hot pressing.

Hot pressing has the advantage of bypassing some flaws occurring in sintered bodies as the applied pressure overpowers the voids in the ceramic, causing them to collapse. Full density is almost assured for fine-starting powders. This is great for ceramic research, but it does not transfer technically to a production sintering process. In these cases where hot pressing is the process of choice, it makes sense to hot press in the lab. Materials that do not sinter readily are candidates for hot pressing. These include refractory carbides, borides, nitrides, some oxides, and composite materials that need a little push to densify. Three subjects: equipment, materials, and procedures will be discussed.

Equipment

Hot presses are electrically heated either by induction or resistance. Commercial equipment is the best choice if one needs a level of sophistication and/or if money is available for purchasing it. Commercial hot presses available for labs are essentially vacuum furnaces with a sealed ram to press on the die.

Induction Heating

An induction furnace was described previously. The hot press differs since the structure has to withstand the stress and to transmit the stress to the die as in Figure 8.29.

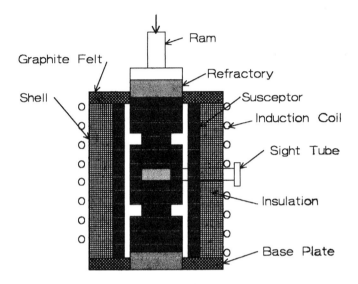

Figure 8.29: Hot Press, Induction. The induction coil heats the susceptor that in turn heats the hot pressing mold.

The ram is part of the hydraulic press. For lab work, a 50-ton press is suitable. When money is scarce, an inexpensive H-frame, hand-pumped press works well. An induced electrical circuit is made through the press frame retrofitted with insulated bolts and sleeves. Refractory blocks are for thermal insulation. Zirconia refractory has low thermal conductivity and is useful for this purpose.

Graphite cylinders transmit the force to the graphite mold. These will oxidize; it is a good idea to include a sleeve for protection. Graphite is machined into a tube for the susceptor. Additionally, thermally insulate the tube with graphite felt or bubble zirconia for use to about 1750 °C. The shell must be an electrical insulator with fiber reinforced cement tubing. Induction coils are copper tubing with circulating water for cooling. Separate the coils with insulated spacers. Match the number of coils to the power supply. The base plate is also insulating and can be a reinforced cement material or refractory. Graphite molds will be discussed further.

Resistance Heating

Many commercial hot presses use a resistance element for heating. One can also use carbon tube furnaces. Figure 8.30 illustrates such a hot press.

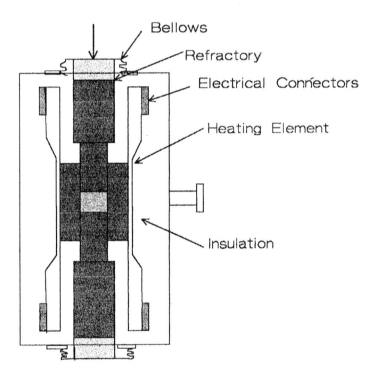

Figure 8.30: Hot Press, Resistance. A tubular heating element is used to heat the mold compressed through end seals with a hydraulic ram.

In this design, the mold is smaller than the heating element. It is easy to separate the mold electrically by wrapping a pressure sensitive adhesive paper tape (masking tape) around the mold on both ends. Then, place it in the press, putting on the end seals that have the bellows, and apply a little pressure. When the paper chars off, suspend the mold away from the heating element. Graphite fiber blanket is the preferred thermal insulation as it is less messy than tamped lamp black.

Optical pyrometers are usually used for temperature measurement. Control the thermal cycle manually or automatically. Due to the short cycles, manual control is perhaps preferable.

Insulated leads (not shown) bring in the electrical connections. A very useful laboratory hot press is one that can be preheated with subsequent mold insertion into the hot zone. This greatly reduces the heating-up time. After pressing, one can withdraw the mold into a cooling chamber. This is shown in Figure 8.31.

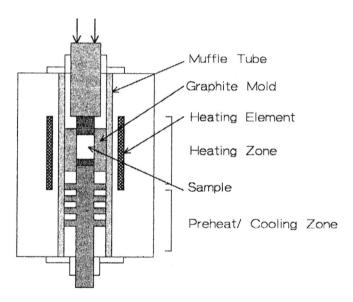

Figure 8.31: Hot Press, Insertion. The hot press is preheated followed by the mold insertion.

Preheat the hot press to or a little above the soak temperature. Then, insert the mold assembly part way into the furnace for preheating just below the temperature that causes the sample to shrink. If it does shrink and pulls away from the walls, the run will be a failure as the sample will fracture and will not reheal. Next, lift the mold into the hot zone and apply pressure. A dial micrometer on the press ram will indicate when shrinkage is complete, which could be as soon as five minutes. Following this, lower the mold below the preheating zone for cooling. Once the mold temperature is below 800 °C, remove the mold, with the total time of about 20 minutes. With this procedure, one can approach the isothermal conditions for hot pressing and can also approximate the densification kinetics. Figure 8.31 does not have some detail of the mold assembly because of the scale. Figure 8.32 shows these details.

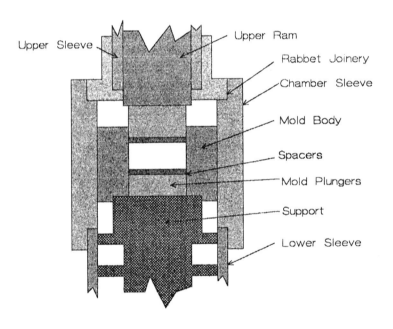

Figure 8.32: Mold Assembly Detail. The mold assembly is joined so that it stays in alignment.

The assembly is supported on the *Lower Sleeve*, which holds the *Chamber Sleeve* in place. The *Upper Sleeve* extends to the top where it is secured, water cooled, and sealed with metal bellows. There is a shoulder (not shown) on the *Upper Ram* to prevent it from dropping out the bottom of the furnace. To load the furnace, the *Support, Mold* with *Spacers*, and *Plungers* are pushed up into the furnace to the desired location for preheating and held there with blocks. The bottom is also water cooled with a copper coil. This furnace was induction heated, also not shown.

Vacuum Hot Presses

As mentioned before, these are essentially vacuum furnaces with the means to introduce a ram for pressing. A metal bellow is a good choice as elastomeric seals wear out. These can be resistance or induction heated and can be heated with metal or graphite heating elements. There are two advantages of vacuum hot presses. One can purge the system, hot press the sample in a relatively pure gas, and process the sample in a vacuum. However, the mold is graphite that can react with the material being hot pressed.

Hot Pressing Molds

Almost all molds are machined from a billet of graphite. ATJ is a fine- grained graphite with good uniformity and strength. Tensile strength increases with temperature from 2200 psi at room temperature to 4000 psi at 2400 °C. Since graphite is highly anisotropic, thermal expansion varies with orientation, being 2.34×10^{-6}/ °C with the grain and 3.46×10^{-6}/ °C across the grain for ATJ. These are low expansions compared with most ceramics, which is fortunate or the hot-pressed part would place the mold in tension when cooled.

Monolithic Molds. Lab molds are usually right circular cylinders with two end plungers. Parts needed for testing are machined from the hot-pressed disc. Parts with an aspect ratio of up to 10:1 can be hot pressed in monolithic molds. The furnace ID, maximum size of ATJ graphite that is a block 9"x20"x24", and the hydraulic press dictate the maximum diameter.

Segmented molds. These have advantages of disassembly. An example is shown in Figure 8.33.

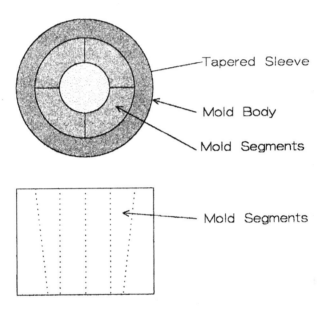

Figure 8.33: Split Mold Design. The interior of a hot-pressing mold can consist of segments that can be pressed out as a unit.

With the segments on a taper, one can easily push out the segments and the part out of the mold and lift the part off. Ordinarily, one can strip the part from the mold, leaving score marks on the ID. These get worse with each run as the interior becomes progressively damaged. With a split mold, the damage is much less, occurring only during pressing. Split molds have another advantage. When the part bonds to the graphite only the segments are sacrificed and the part is more easily recovered. A thin sleeve of graphite, graphite foil, or Mo foil can line the ID, preserving the segments. A split mold is more expensive due to the extra machining, but this is not a big factor in lab work, and the mold body can accommodate a variety of segment ID configurations. The following Table 8.1 lists a few thermal expansions of common materials.

Table 8.1: Thermal Expansion of Common Materials

Material	Thermal Expansion % to 1000 °C
SiO_2 ,glass	0.055
Si_3N_4, (beta)	0.22
Graphite,ATJ	0.23 with grain, 0.35 across grain
Si_3N_4 (alpha)	0.28
WC	0.48
SiC	0.5
B_4C	0.54
ZrO_2 (monoclinic)	0.74
TiC	0.75
Al_2O_3 (alpha)	0.83
Material	Thermal Expansion % to 1000 °C
$MoSi_2$	0.83
BN	1.3 (anisotropic)
MgO	1.33

Any material with a lower thermal expansion than graphite will jam in the mold as it cools. Since the part will shrink less than the graphite, there will be an interference fit on the interface, placing the part in compression and the mold in tension. Fortunately, there are not many materials with this low an expansion. As one approaches the expansion of graphite, one can expect greater difficulty in stripping the part. The split mold bypasses this problem.

Hot Pressing Tubes. The preceding conclusions on thermal expansion referred to the stresses on the ID of the mold, but what if the intention is to hot press a tube? In this case, the ID of the part is in tension and the OD of the central graphite form is in compression. The part will break for sure. Figure 8.34 represents a partial solution.

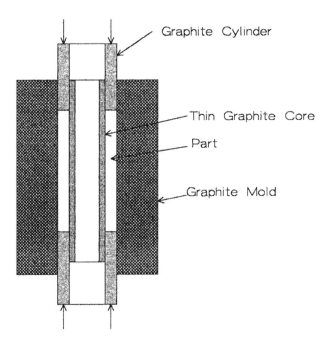

Figure 8.34: Mold for Tubular Shape. The central tube has to be thin enough so that the part does not fracture in tension during cooling.

The central graphite core must be thin walled in order for it to yield or collapse during cooling. If it is too thin, the wall will collapse during hot pressing, so it has to be just right. This might take a few trials. Aspect ratios are about the same for tubes as for cylinders, but the pressure has to be limited, especially at first. When the plungers start to move, the material in the mold is deformable to some extent, and the graphite gets a little stronger. One can now gradually increase the pressure.

Cones. One can hot press conical shapes as the sintering ceramic has some capacity for lateral movement. Not much, but some. Figure 8.35 sketches a mold where the top plunger is in two parts to compensate for the longer vertical movement of the thicker section.

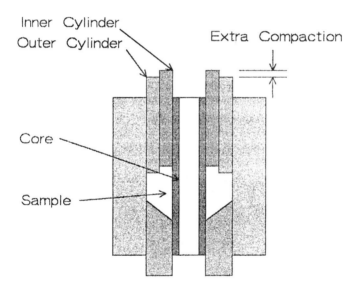

Figure 8.35: Mold for Conical Shape. Due to the large difference in height, an extra pressing ring may be needed to obtain full densification in the center.

Reactions with the Graphite. Materials that require a high temperature to densify will react with the mold and will tend to stick. This is especially true with carbides, borides, and nitrides. In the worst case, it may only be possible to recover the part by sand blasting the mold. Sleeved segmented molds are useful here if the temperature is low enough for Mo foil (about 1700 °C). Graphite spacers and liners can simplify recovery of the part. A mold of this type is sketched in Figure 8.36.

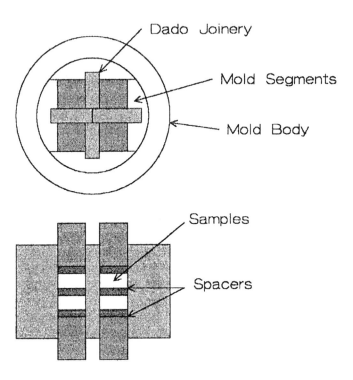

Figure 8.36: Mold with Spacers and Liners. A multiplicity of cavities can be made with spacers and dividers.

This mold hot presses eight pieces per run. Position the graphite dividers with dado joinery. The entire core assembly is pushed out and the parts recovered by disassembly. It is practical to use segmented molds when one needs many identical parts. Segmented parts are in contact with the graphite and require selection to avoid unwanted reactions.

Triaxial Hot Press Design. All of the preceding molds have a monolithic shell holding the parts together. One can transfer this holding restraint outside the furnace with push rods. Figure 8.37 is of a view looking down into the furnace.

Figure 8.37: Triaxial Hot Press. Interior mold parts can be held in place with rams, placing all of the ceramic mold parts in compression.

The mold is in the center and consists of four segments restrained with four push rods that hold them in place. Two of these rods butt up against a square steel frame. The other two rods (at 90 degrees) are connected to hydraulic cylinders that supply the thrust holding the mold parts in place. The plungers that compact the sample come out the top and connect to a hydraulic press. In the above figure, the top was removed to expose the interior construction. The mold in Figure 8.37 makes a cylindrical sample. Figure 8.38 is of another mold with a square cross section.

Figure 8.38: Triaxial Mold Parts. The four parts are dissembled after the sample is hot pressed. They can be lined with Mo foil to prevent reactions.

The sides of the mold are identical in size and shape and fit together with rabbet joinery. The punches are rectangular blocks. While called a *triaxial* design, pressing is uniaxial with the other two axes only holding the mold parts in place. The inherent advantages to this design are as follows.

- All ceramic parts are in compression.
- A variety of ceramics can be used for the mold.
- Higher pressures can be used for hot pressing.
- There is little graphite contamination.
- Mo foil spacers are used on the walls and ends.
- The sample is recovered by disassembly.
- Triaxial designs can be scaled up.

One can make molds from recrystallized silicon carbide, titanium diboride, or other refractory materials. Push rods can be made from recrystallized silicon carbide. Silicon carbide has lower thermal conductivity than graphite and is stronger; therefore, the rod diameter is smaller. This lowers the heat loss out through the push rod.

Materials

Hot pressing is more expensive than sintering and substitutes for sintering to densify the part or to produce the necessary superior properties. Some examples will be discussed.

Oxides

There are two approaches to obtaining a very fine microstructure. Hot pressing: where pressure reduces the temperature and limits grain growth. Alternatively, very finely-divided, high-purity powders that sinter at lower temperatures can be used. The trend is toward the latter approach, but hot pressing reduces the flaw population to 10-μm cavities or larger cavities. Alumina cutting tools have been hot pressed to advantage.

Carbides/ Borides/ Nitrides

These diamond-like structures are difficult to sinter and are often hot pressed. Without use of boron and carbon additives, SiC does not densify during sintering.[5] Otherwise, one needs to hot press SiC. Sintering and hot pressing temperatures are over 2000 °C; this adds to the difficulties and cost. Hot pressing boron carbide (B_4C) yields a high density and the best properties. Sand blast nozzles and other wear parts are made from hot pressed B_4C. TiC and TiN are often hot pressed. Si_3N_4 is usually sintered with alumina and yttria sintering aids.[6] These combine with silica in the material to form an interstitial glassy phase. This may not be the best solution for making high quality parts, but many shapes are not conducive to hot pressing technology.

Composites

During sintering, it is not uncommon for one of the phases to interfere with the densification of the other. This can be overcome by force in the hot press. Maybe the best example is alumina-cutting tools reinforced with silicon carbide whiskers. Whiskers form a tangle that gets in the way of alumina sintering. These tools are very successful due to their abrasion resistance and especially their toughness. Unlike sintering, one can hot press metal inserts into ceramic structures. A good example is shown in Figure 8.39 where steel fasteners are molded into an alumina ceramic.

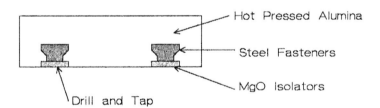

Figure 8.39: Hot Pressed Ceramic with Metal Fasteners. Metal inserts can be integrally hot pressed into the ceramic and used as fasteners.

Hot press the alumina at a low enough temperature to avoid reactions with the steel. This can be done by using a submicron powder, a glass interstitial phase, or a sintering aid such as titania. Steel inserts are set on MgO pressed powder discs to isolate the steel from the graphite mold and to prevent their reaction. MgO does not sinter at low temperatures and is easily removed after hot pressing. Since steel (iron) has a higher thermal expansion than alumina, the insert is loose after cooling, but it is locked in place by its shape. One can drill and tap the steel inserts making the composite suitable for bolting to a structure. Figure 8.40 shows another example of a hot-pressed composite. It shows the placement of a fine-grained facing on a coarse-grained backing.

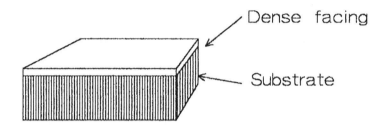

Figure 8.40: Hot Pressed Ceramic with Dense Facing. Two different materials can be hot pressed together. Here, a dense facing was pressed on a porous backing.

By facing an alumina refractory block with a dense, fine-grained, alumina ceramic, the surface becomes impervious and wear resistant. This could not be done by sintering because of the huge difference in shrinkage during firing. This is not a problem in hot pressing. Another technique is to mold a metal reinforcement into a ceramic structure, seen in Figure 8.41.

The figure is schematic as there are limits on the amount of Mo metal that can be successfully incorporated because of the lower thermal expansion. One example is wire or perforated sheet in hot pressed alumina. Alumina penetrates the spaces between the metal during hot pressing and forms a composite structure. This would not be possible with sintering since the ceramic shrinks and the metal does not. Interestingly, boron has a thermal expansion very close to alumina up to 1000 °C and boron fibers have been made. This could make an interesting composite.

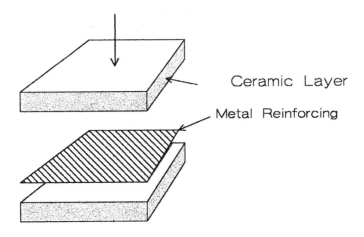

Figure 8.41: Hot Pressed Ceramic with Metal Reinforcing. A Mo mesh can be hot pressed in a ceramic to provide reinforcement.

While not a composite, there is another interesting thing that one can do with hot pressing. Coin a preform out of graphite powder with a particular surface contour, which will be reproduced on the ceramic surface. Look at Figure 8.42.

Reproduce the surface of the penny by placing it in a die, pour in a fine graphite powder, and press. Place the graphite form in a hot pressing die, pour in fine alumina powder and level it. Recover the alumina reproduction after hot pressing. While this is not a good way to make money, the technique could be useful for more reasonable things.

Figure 8.42: Coined Alumina Penny. By pressing graphite powder over a contoured face, a reproduction can be made with a ceramic part.

Check List, Hot Pressing

- Decision to hot press
- Equipment choices: resistance, induction, vacuum, insertion, triaxial
- Equipment design
- Molds: graphite type, monolithic, segmented, foil separators
- Special cases: low expansion ceramics, tubes, cones
- Reactions with graphite
- Ceramic materials: oxides, carbides, nitrides, borides
- Composites: matrix+whiskers, metal embedded fasteners, metal reinforcements, hard facings
- Coining

6.0 HIPPING

A HIP is a hot isostatic press that was developed at Battelle and later commercialized. This equipment is commercially available so do not try to build one. Gases under pressure are dangerous. One refers to the pressure vessel as a bomb, and for good reason. Commercial HIPs are safe and no explosion has ever occurred. When they fail, it is due to a leaky seal, which is not dangerous.

Equipment

Figure 8.43 depicts a large HIP apparatus.

Figure 8.43: Hot Isostatic Press, Commercial. HIPs of various sizes, temperature, and pressures are available. (Courtesy of ABB Autoclave)

While this is a large HIP, the features in a lab size are similar. There are at least two reasons why a small HIP is not a good investment. The next job may require a larger ID, a small HIP does not have enough clearance in the heater, and deposits can short out the lead throughways. For most lab work, an ID of 4 inches to 6 inches is reasonable.

Figure 8.44 is a sketch showing the various features of a HIP.

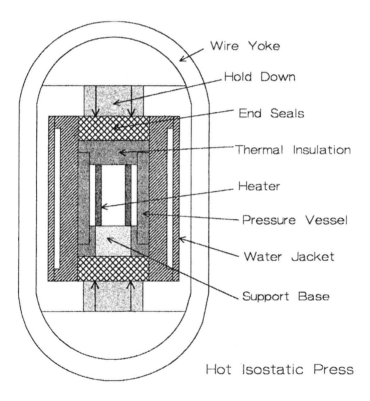

Figure 8.44: Hot Isostatic Press, Sketch. The various parts of a HIP apparatus are shown in the sketch.

Starting from the outside and working one's way in, the press end closures are secured with a wire-wound yoke. Wire-wound structures are tougher than solid steel, which is why they are used for suspension bridges. Some HIPs have breach block end closures that also work well. Pressure vessels are heat-treated, alloy steel with a good safety factor, and they are pressure tested above the operating specifications. The vessel is water cooled with a jacket. For graphite heaters, the thermal insulation is graphite fiber molded into shapes. Slotted graphite heating elements are common. The base of the heater is a structure that supports the sample and element. It also connects to the power supply. When the heating element is a refractory metal, the thermal insulators are radiant heat shields. The internal structure is similar to that in vacuum furnaces. When using metal construction, the atmosphere is a lot cleaner. Graphite turns a white sample dark grey. Temperature is sensed with a thermocouple. Pt-Pt/Rh is preferable if the temperature does not exceed 1650 °C, otherwise use W/Re.

Enclosures

It is common practice to isolate HIPs for safety reasons. When cramped for space, surround the HIP with sliding panels that run on both bottom tracks and hang from rails supported by stanchions. Panels are made to manufacturer's specifications, 1/4-inch steel with 2-inch plywood fastened on the HIP side of the panels. Plywood traps the floating shrapnel, while steel prevents penetration, not that this is likely to occur. A big advantage is that by sliding the panels around, access is available to all sides of the apparatus.

Difficulties

Some HIP designs result in difficulties, which is of interest as it is a warning to the pitfalls in design that should be avoided.

Meltdown

The base support has only a little clearance around the electric lead throughways. Deposits can condense in this area, shorting out the base unit, which will arc and destroy the furnace. This equipment is very expensive, and one can lose much time. Do not buy the smallest unit. Get one large enough to provide ample clearances, and keep them clean.

Jamming

Hold-downs are pressed onto the end closure with steel rods that have a ball bearing riding on the top of the closure. Bearing races can be under-designed and will crush. The yoke cannot be rolled back, preventing the apparatus from opening. The bottom end closure also pushes out and can jam against the yoke extension. There is something indecent about beating on an expensive laboratory instrument with a sledgehammer, especially when repeated after every run.

Procedures

There are three basic procedures when running a HIP based on the method of isolating the material from the pressurized gas. Hipping only works if there is a pressure differential between the pressure vessel and the sample. This differential provides the stress that aids in the densification process. Provide isolation in three ways. Encase the sample in a refractory metal (Mo) can, sinter the sample to an impervious condition, and then hip it. Alternatively, encase the sample in a silica glass that fuses to form an impervious barrier.

Canning

This was the first method. Place the sample in a sheet metal container and then weld it closed. The metal is either Mo or Ta, both of which require electron beam welding in an inert atmosphere. Then, place the can in the HIP, where the heating/pressure/time cycle is run. The can collapses as it is pressurized, transferring the stress to the sample. Special shapes such as hollow cones can be hipped against a central form that forms an accurate ID.

Sinter HIP

In this method, the sample is first sintered to an impervious state and then hipped. One usually sinters and hips in the same run, but while being convenient, this is not really necessary. After excluding the internal porosity from the pressure, the pressure hips the sample. This process is the easiest one to use under normal conditions of densifying a part. A prerequisite is that the material will densify to an impervious state.

Glass Encasement

At first, the process was to submerge the sample in a bath of silica glass and then to pressurize it. In a later process, the sample is coated with the glass, heated to melt the glass, and is then pressurized. Since the silica glass has a very low thermal expansion, it cracks upon cooling and spalls off. This process developed by ASEA requires a license even for experimental purposes. This requirement makes it difficult for a lab investigation.

Check List, HIP

- Do not buy a small HIP, buy at least a medium-sized unit.
- Keep the lead throughways clean.
- Consider jamming. Call users who have experience with the equipment.
- Sinter HIP is the most useful process, generally.
- Glass encapsulation is a production process and is not available for experimentation.
- Metal encapsulation is expensive, but is useful in special cases.

REFERENCES

1. J.H. Heasley, Internal Report
2. A.H. Heuer, "Alloy Design in Partially Stabilized Zirconia," in <u>Science and Technology of Ziriconia 1981</u> Ed. A.H. Heuer and L.W. Hobbs, (Westerville, OH: Amer. Ceram. Soc., 1981), pp.98-115.
3. John I. Fredriksson, Slip Cast Silicon Carbide Shapes. U.S. Patent 2,964,823 (Dec. 20 1960).
4.   J.H. Heasley, and A.G. King, Internal Reports
5. Prochazka, Svante, U.S. Patent 4,004,934 (1977).

9

Grinding

1.0 THE GRINDING PROCESS

The scope of this chapter is limited. Grinding is by itself a skilled or semiskilled craft. The intent is not to teach how to grind a ceramic part. In-stead, the intent is to describe the process and offer some insight about how this affects the ceramic.

Ceramics are usually ground with bonded-diamond abrasives. There are two types of bonds in common use, metal and resinoid. Metal is rougher and lasts longer. Resinoid is gentler and wears out faster. Always use resinoid for finish grinding of fine-grained, dense ceramics as there is less surface damage to the part.

Tool Type and Grit Size

There are two types of metal-bonded, diamond-cutting tools. The sintered-powder, metallurgical type with included diamond grit and the surface bonded type with electroplated Ni on sedimented diamond grit. The latter is common in metallographic grinding and polishing. The powder-metal, bonded type is more common in the machine shop and is used for percision shaping of a part. Resin-bonded and diamond-bonded cutting tools also come in these two forms for similar uses. Abrasive points protrude from the surface of the grinding wheel, as schematically shown in Figure 9.1.

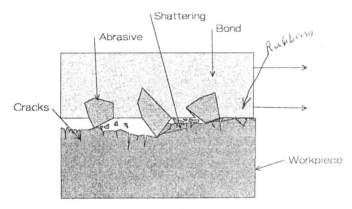

Figure 9.1: The Grinding Process. Abrasives damage the ceramic surface to an appreciable depth. Rubbing of the wheel on the surface can cause heating.

Grinding is a scratching process, leaving jagged scratch roots and sub surface cracks. In Figure 9.1, not all the abrasive grains stick out from the wheel surface the same distance, so the longest grains do most of the grinding. Try polishing a finely ground surface. One will observe resulting deep scratches on the surface. Grinding also leaves residual stresses in the ceramic surface and a considerable amount of plastic deformation.[1]

Grit size is a factor when grinding ceramics. Finer grit leaves smaller scratches and lower residual stresses as long as the grinding wheel is free cutting. The down feed has to be less than the distance of the protruding diamond points from the surface. Wheels will dull with use and have to be dressed to free up a fresh cutting surface. Since these are diamond wheels and are expensive, there is a reticence to dressing the wheel and it is often abused when dull. Dull wheel surfaces rub and can raise the interfacial temperature even to the point where the part could fracture. In Figure 9.1, rubbing is shown to the right of the contact area. Pullouts leave the resin bond in contact with the work piece, generating heat but without any cutting. There is proportionally less cutting and proportionally more rubbing with a finer grit wheel. There is then a practical lower limit to the grit size that usually is 220-325 grit.

Another limitation for most shop surface grinders is a minimum down feed of 0.0001 inches; this is not too precise especially after adding in all of the machine tool compliance. With a small grit size, the distance the grain protrudes is of the same order as the down feed, jamming the wheel surface against the work piece. The modulus of many ceramics is about $30\text{-}60\times10^6$ psi while the modulus of phenolic resins is about 1×10^6 psi. Forces between the cutting point and the ceramic surface can push the point up into the wheel surface either elastically or by deformation of the bond. All of this is why there is a practical lower limit on the grain size that is effective. When one needs a finer surface, switch to lapping or polishing with loose abrasives. The part wheel interface in Figure 9.2 is at a higher magnification.

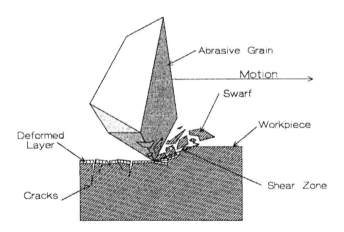

Figure 9.2: Grinding Damage. An abrasive grain scratches the ceramic surface. Damage includes: gouges, cracks, and surface deformation.

The above figure shows an abrasive grain scratching the ceramic surface. This leaves damage in the ceramic, including gouges, cracks, and a deformed surface layer. The grain creates a shearing zone in front of the cutting tip. Fracture occurs in this shearing zone, removing material.

Surface Damage

The grinding process damages the ceramic surface. Scratches, because of their shape, act as stress risers. Sub surface cracks and residual stresses weaken the ceramic. Two cases are illustrative, lowering the modulus of rupture and lowering the impact strength.

Modulus of Rupture

This example is taken from work on high purity, hot pressed alumina, fully dense, and with an average grain size of three μm.[2] Chemically polishing alumina with a borax fusion allows removal of the damaged layers. Figure 9.3 illustrates the solution rate of strength vs. depth of damage.

Strength increases as material dissolves off the damaged surface, from 72,000 psi to 95,000 psi over a depth of five μm. This depth is also about the size of the largest grains in the microstructure. This leads to the unproven but likely conclusion that the damage is blocked at the grain boundaries.

Impact Strength[3]

Using the same type of ceramic as in the preceding section and varying the hot pressing temperature results in a range of grain sizes. All of the specimens were fully dense. With the discs supported by an annular knife edge around the edge, a half-inch steel ball bearing was dropped onto the center of each disc at a variety of heights. Figure 9.4 illustrates the results.

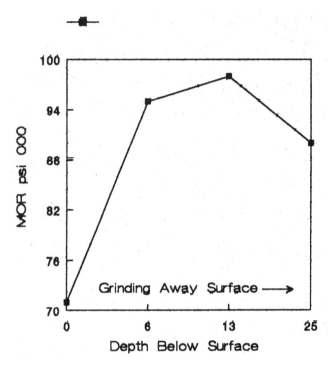

Figure 9.3: Strength with Distance from the Surface. Strength increases as the surface damage is dissolved away.

There are three important observations in Figure 9.4. Impact strength decreases markedly as one reduces the hot pressing temperature, coarser grit grinding lowers the impact strength, and annealing at 1300 °C removes the grinding damage. (By way of explanation, the tools were not available in the lab to measure the grain submicron size, so temperature was the criterion by default.) Annealing is not a common practice for fine-grained, dense, ceramic parts. Maybe it should be tried more often, unless this results in bloating.

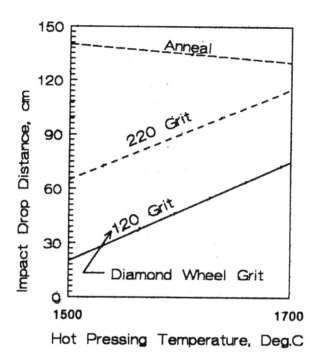

Figure 9.4: Impact Strength/ Microstructure. Impact strength is substantially reduced with lower hot pressing temperatures and larger grit grinding wheels. Damage can be removed by annealing.

2.0 GRINDING SHOPS

There are three-setup, grinding choices: another department in the institution, a setup in the ceramic lab, and outside sources. It is well worth the time to consider these alternatives.

Another Department

Machine shops in the institution are usually in another department, which have their own priorities. While also doing work for production departments, lab work is at the bottom of the list, and sometimes for all practical purposes, not even on the list. The process for making parts in the lab is sequential: processing powders, forming, sintering, and shaping. Each of these steps may require machine shop services. Consider the worst scenario that one may encounter. For example, a filter may need to be adapted for powder processing, a die may be needed for pressing the part, an end seal may be required for atmosphere sintering, and test specimens have to be ground for measurements. With a four-week hiatus between each step due to machine shop scheduling, it will take too long to get the experiment completed. The author retracts an earlier statement; this is not the worst scenario as a four-week queue is quite timely for such jobs.

Availability for machine work varies depending on the location in question. Generally, place the choice of an inside department on the bottom of one's list.

Setting Up Grinding in the Ceramic Department

When under control of the lab, priorities are set internally, which can be a great relief. However, there is the problem of getting the necessary skills, and this can be a sizable obstacle. When the work in question is repetitive, such as grinding test bars, the level of skill is manageable. A comprehensive grinding shop requires a lot of expensive equipment and knowhow. With operations going broke, both equipment and skilled people are sometimes available. This will cut the lead time, cost, and acquisition of skills. There is also the question of facility utilization. It will take much work to fill the capacity of a comprehensive shop. When these requirements are well satisfied, an internal machine shop could well be the preferred choice.

Outside Contractors

One can contract work to a variety of preexisting machine shops that also machine ceramics; these machine shops do advertise.[4] Metal working shops are plentiful. This is advantageous as skills or capabilities can be selected; there are very talented people who will be delighted to do business. Another big advantage is that shops do not get paid until completion of the work and to specification. It is surprising how efficiency is promoted when money is involved. Perhaps, the main virtues of contracting with outside shops are the skills, experience, and delivery time. For such a contract, one recommends establishing a good rapport with a free exchange of information.

3.0 TYPES OF GRINDING

There is no intent to make this section comprehensive, as that is beyond the scope of this discussion. The intent is for general familiarization and some specialized comments that may be helpful. First under consideration will be a summary of grinding methods.

Grinding Methods

There are many types of grinding machine operations. The most common grinding methods for ceramics are as follows: surface, cylindrical, cutoff, universal, belt, and lap grinding. There are a variety of attachments that can be added to the grinding machine for grinding, such as IDs, faces, slots, holes, and chamfers; these will be briefly discussed.

Surface Grinding

These machines have three slides: vertical, cross, and lengthwise. There are several ways to secure the parts that need to be ground. Magnetic chucks are useful but if the ceramic is not magnetic, one will need to glue the parts to a steel plate that is ground flat and parallel. Hot melt adhesives have excellent holding power and are available as sticks at any hardware store. Place the ground steel plate on a hot plate and rub the stick to melt the adhesive onto the surface. Preheat the parts on the hot plate and press them onto the adhesive with a twisting motion to seat them snug. As everything is blistering hot, use a tool. After grinding, reheat the assembly and remove the part. Again use a tool to prevent burns. The adhesive is soluble in trichloroethylene, which is a safew, non ignitable solvent. Trichloroethylene is somewhat toxic so one should use a hood or vent to prevent vapor inhalation. Waxes are commonly used for cementing parts, but waxes are not as good as hot melt adhesives for securing the part. With waxes, the part may come loose with resulting damage.

Another way to mount the parts is with steel blocks placed on the sides and held in place with the magnetic chuck. These can come loose, so care is required for safety.

One usually uses coolants when grinding ceramics. The supplier of the grinding wheels will recommend the coolant suitable for this application. The coolant is held in a sump that has a baffle, allowing the swarf to settle out. It is pumped with a nozzle directly onto the sample surface just in front of the wheel contact. Replace the coolant and clean out the sump regularly, depending upon the amount of use.

There are a variety of grinding wheel sizes and shapes, each of which is preferred for the job in hand. Besides configuration, the wheel has specifications on grit size, bond, and concentration. One uses coarse grits, such as 120 size grits, to remove most of the stock. One uses finer sizes, such as 220-325, for finishing. For fine-grained, dense ceramics, the preferred bond is resinoid. Coarse, friable ceramics, such as some refractories, are so abrasive that a metal bond is necessary because of its better wear resistance. These often use grit sizes on the coarser end.

There are two principle types of diamond: natural and synthetic. Rough grinding applications usually use natural, as it is more durable.

Synthetic diamond grit is almost exclusively used for fine ceramics as it is more friable and freer cutting. It is available in a few grades based on friability. Additionally, synthetic is quite a bit less expensive than natural diamond.

Concentration numbers range from 50% to 100% representing the amount of diamond in the bond, with 100% being the maximum amount of diamond that can be successfully incorporated. Fine ceramics use 100% concentration. Figure 9.5 is of a surface grinder designed especially for ceramics.

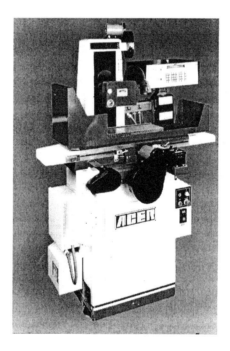

Figure 9.5: Surface Grinder. Grinding machines should be rigid and dampen vibrations. (Courtesy of Chand Kare)

The principle differences between this machine tool and a metal working machine pertain to the parameters relevant to grinding ceramics. These are CNC controlled and include wheel speed, table motions, depth of cut, and coolant flow.

Ceramics are more sensitive to surface damage than metals because ceramics are brittle while metals are usually ductile. Additionally, almost all ceramics are thermal insulators whereas almost all metals are thermal conductors. This results in the contact surface temperature, for a ceramic grinding, running hotter than for a metal. The machine, in Figure 9.5, has a cast iron base that is more rigid than a weldment structure. Table motion can be either servomotor driven or the more conventional hydraulic system. Grinding wheels are a critical part of the grinding process. One important wheel factor is the distribution of abrasives in the wheel bond. It turns out that it is difficult to evenly disperse a coarse-grained material in a fine-grained, metal, or polymer powder. The two materials tend to segregate when handled.[5]

The coolant is selected and filtered for the particular ceramic being ground. Now, how are all these parameters chosen for a particular situation? Part of the answer is by experience and part by measuring the *grindability*. With grinding metals the "G" ratio is often used. This is an efficiency factor relating to the ratio of wheel wear to the amount of metal removed. However, for ceramics, the "G" ratio is so high due to the use of diamond wheels that it is not useful in this case. Instead, one way is to measure a *grindablity* factor. This factor relates the material removal rate to the wheel speed and the normal force between the abrasive tool and the work piece. Figure 9.6 illustrates this test stand.

Testing proceeds by forcing the end of a standard (3x4x45 mm) test bar against a diamond-grit-coated belt with a coolant. Adjustable weights determine the normal force. Use a fresh position on the belt for each measurement. Belt speed is adjustable. One measures the change in length of the test bar with time. A general guideline is that with low grindability ceramics, grinding becomes more difficult, advising slower wheel speeds, less down feed, free cutting wheel specifications, ample coolant of the appropriate type, and slower transverse table motion. In other words, take it easy.

Figure 9.6: Grindability Test System. The test results help in selection of wheel specifications and grinding conditions. (Courtesy of Chand Kare)

Cylindrical Grinding

A simple cylindrical grinder merely rotates the part, with limited tooling options. One can grind ODs and IDs. Inside diameters have special problems in that the geometry imposes stresses on the ceramic surface. The wheel diameter is usually small, requiring a high-speed, internally-grinding spindle. For long ID grinding, deflection is more critical because of the longer lever arm holding the wheel. Depending on the tolerance factor, one may need to make the setup as rigid as possible.

CutOff

To minimize the volume of removable material and to increase the cutting speed, the diamond blades for cutoff purposes are usually thin. The blade thickness for lab use is often 0.035 inches. Unit pressures are high, making resin bonds short lived. One uses resin bonds to minimize damage to the ceramic, but ordinarily one uses metal bonds. Only the rim of the disc contains the diamond abrasive that is easily damaged when bent or when a piece breaks out. The diamond-containing rim of the blade is wider than the steel disc, providing clearance so that the disc clears the sides of the cut in the ceramic. When the cutoff blade enters the workpiece at an angle, it is deflected down the slope and the steel disc will rub. Also, the side of the diamond-containing rim will rub and wear. When the *set* of the rim is gone, the blade will not clear the sides of the cut and for all practical purposes the blade will be prematurely worn out. Encounter between the blade and ceramic surface should be at 90 degrees, and when it is not, a small notch should be cut where the blade is to enter the work so that it will not wander off square.

One can use cutoff blades on a surface grinder or on a cutoff machine. Many lab machines use an 8-inch diameter blade. On the surface grinder, one can gang up the blades with spacers between each blade. This is especially useful when grinding many identical parts, like test bars. These discs are smaller, where 4-inch D and 0.020-inch thickness is common.

Figure 9.7 is of a CNC-controlled, cutoff machine that either uses single or ganged cutoff blades with controlled wheel speed and feed rate.

Universal Grinders

These are versatile grinding machines that have more than one spindle and the capability of multiple setups. These machines are very expensive and not commonly found in the lab. One may need such machines for some types of complicated or high precision work. Some manufacturers include air bearings and vibration absorbing bases with multiple (3) spindles.

Figure 9.7: Cutoff/Gang Slicer. Multiple cuts with ganged wheels speed up the cutoff process. (Courtesy of Chand Kare)

Belt Grinders

The belts are essentially sanding belts such as those found in wood working. Diamond abrasives are available but not common. The problem with belt grinding on ceramics is that the pressure between the belt and ceramic is limited, resulting in a low rate of cut. It is not practical to use belts on hard, dense ceramics, except perhaps for polishing a curved surface where the belt conforms the contour.

Laps

One uses laps for superfinishing or polishing. Superfinishing calls for a loose abrasive in oil on a metal lap. While the surface shows good readings with a profilometer, it retains scratches. Polishing is done on a fabric lap surface and uses a variety of abrasives for different applications. The combination of fabric and abrasive is often specific for each material. Table 9.1 lists some examples.

Table 9.1: A Variety of Fabrics, Abrasives, and Coolants for Different Materials

Material	Fabric	Abrasive	Coolant
Alumina	Nylon	Diamond	Oil
Zirconia	Texmet[6]	Diamond	Oil
Glass	Felt	CeO_2	Water
Plastic	Felt	Al_2O_3	Water

The Ceria and Alumina polishes are obtained from ophthalmic supply houses.

4.0 OTHER MACHINING METHODS

Several of these machining methods will be briefly mentioned.

Ultrasonic

This method places an ultrasonic probe just above the part surface, in a water bath containing an abrasive. The abrasive vibrates, impinging on the surface and nibbling it away. Hole shape is that of the tool, making this method especially useful for odd shapes. Tool material is often a soft metal (often copper) that resists abrasion. Soft metals tend to smear rather than cut and can be durable under these conditions.

Electro Discharge Machining (EDM)

Tooling for EDM looks similar to that used for ultrasonic in that there is a shaped probe just above the surface of the ceramic. With the ceramic grounded and the probe charged, an electric spark erodes the ceramic surface. For this to happen, the ceramic has to be an electrical conductor, placing a restriction on where the technique is applicable.

Check List, Grinding

- Grinding Shop
    Departmental
    In the lab
    Outside sources
- Grinding Method
    Surface
    Cylindrical
    Cutoff
    Universal
    Belts
    Laps
- Other Methods
    Ultrasonic
    EDM

REFERENCES

1.   D. Johnson-Walls, A. G. Evans, D. B. Marshall, and M. R. James, "Residual Stresses in Machined Ceramic Surfaces," *J. Am. Ceramics Soc.* 69[1] 44-47 (1986).

2.   A. G. King, "Chemical Polish and Strength of Alumina," Materials Science Research Vol.3, (Plenum Press, 1966), pp. 529-38.

3.   Alan G. King and W. M. Wheildon. Ceramics in Machining Processes. New York: Academic Press, 1966, p 63.

4.   *American Ceramic Society Bulletin*, Westerville, Ohio.

5.   Roland H. Chand, Personal Communication.

6.   Buehler trade name.

10

Effects of Processing on Properties

1.0 INTRODUCTION

Crafting a ceramic is similar to building a pyramid. It starts with a base, which, in this case, is the starting grain or powders. Everything depends on the base, and when it is insecure, the pyramid will not stand.

Both microstructure and superior properties derive from the starting materials, provided that these materials are handled with respect throughout the remaining processing. Preceding chapters addressed how careful handling is accomplished. Remaining to be addressed is the important task of understanding the results.

This chapter describes how materials selection and processing affects the properties of the final ceramic.

2.0 SELECTION OF MATERIALS

The characteristics of the starting materials are crucial and complex: there are many alternatives that depend on what is to be accomplished. Addressed below are three important factors: physical

333

properties of the phases, chemical properties of the phases, and the final microstructure of results from the materials selected at the beginning. These preceding factors serve as guiding restraints; the final selection comes from experience.

Physical Properties of the Phases

Different materials have unique intrinsic properties that derive from the chemical bonding and crystal structure of the lattice. The initial choice of powdered and granular materials depend upon these properties.

Melting Point

The melting point of the phase limits the material's temperature applications.

Thermal Conductivity

High conductivity reduces thermal shock and conducts heat. Materials with low thermal conductivity are useful thermal insulators.

Thermal Expansion

High expansion increases the tendency for thermal shock to occur.

Heat Capacity

High heat capacity retards heating and cooling.

Electronic Conductivity

Different applications require either conductors or insulators.

Ionic Conductors

Such conductors are used in oxygen sensors and fuel cells.

Hardness

As a general rule, hard materials are more wear resistant. However, exceptions to this rule do exist.

Chemical Properties of the Phases

Chemical properties determine which phase will be stable at a particular temperature as well as the reactivity of a given application.

Thermodynamics

The free energy of formation predicts reactions occurring at elevated temperatures. Thermodynamics also predicts the phases that are stable at different temperatures. Many phase diagrams exist and are useful aids when selecting the starting raw materials.

Solubility

A material can dissolve when it contacts a liquid. Because most phase diagrams do not provide solubility information, a material's solubility must often be determined experimentally.

Impurities

Impurities in the starting materials can greatly affect the final ceramic. So much as cost is not a limiting factor, the use of clean materials is much preferred.

Microstructure

Of concern is the effect of the starting materials on final microstructure after sintering. There do exist fine and coarse-grained materials. However, other considerations include sinterability, final grain size, shrinkage, porosity, and flaw populations. These properties are often characteristics of the starting materials.

Sinterability

Very fine powders sinter at lower temperatures and produce a finer grain intercept. Sinterability is also influenced by particle shape. Irregular particle shapes result in excess pore coalescence and a decreased sintered density. Blocky shapes pack well and sinter in an orderly fashion.

Firing Shrinkage

Fine powders exhibit higher firing shrinkages for the following two reasons: these powders sinter to higher densities, and as the particle size becomes increasingly fine, surface forces between particles prevent packing to a high green density and result in a larger volume change upon densification.

Flaw Populations

Flaws in the green body greatly influence the fired strength. Strength increases and strength distribution narrows as flaws in the starting powder are removed. This effect is illustrated in Figure 10.1 as adapted from the literature.[1]

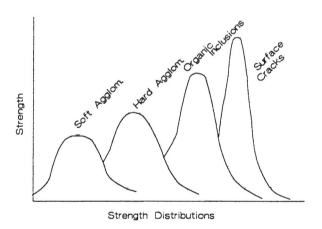

Figure 10.1: Strength Effect of Removing Flaws. Strength increases as flaws are removed from the starting powder and surface grinding.

While the figure is schematic, it illustrates the strength behavior as flaw populations are eliminated: strength increases and the distribution narrows. Two options for removing flaws are to select a better powder or clean up the existing powder. Cleanup is accomplished by milling, leaching, classifying, and calcining.

Agglomerates in the starting powder often result in flaws in the fired body, an example of which is shown in Figure 10.2. Fracture origins are usually seen on fracture surfaces for fine-grained ceramics. This origin is an area where grain growth occurred, probably due to an impurity.

The relatively smooth area around the inclusion is called the mirror. In a homogenous material such as glass, the mirror, like its common-day counterpart, can be featureless. The mirrors on fine-grained polycrystalline ceramic fracture surfaces are less distinct, but still observable. Coarse-grained ceramics may not exhibit a fracture origin due to their heterogeneity. Most fracture surfaces exhibit a crack origin with the accompanying mirror. Sometimes these origins are inclusions, as depicted previously, but they can also be voids, areas of coarse grain size, or surface flaws caused by machining.

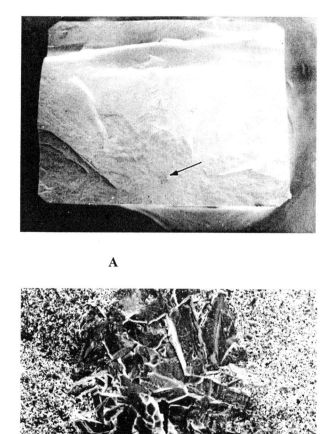

A

B

Figure 10.2: Fracture origin. **A** scale bar 1000 μm, **B** scale bar 100 μm.

Check List, Selection of Materials

- Physical Properties
    - Melting point
    - Thermal conductivity
    - Thermal expansion
    - Heat capacity
    - Electronic conductivity
    - Ionic conductivity
    - Hardness
- Chemical Properties
    - Thermodynamics
    - Solubility
    - Impurities
- Microstructure
    - Sinterability
    - Firing shrinkage
    - Flaw populations
    - Clean up of flaw populations

3.0   EFFECTS OF TEMPERATURE AND PRESSURE ON PROPERTIES

Since temperature and pressure often interact, these properties are discussed together below.

Density, Die Pressed TZP

Figure 10.3, presented as a contour map, shows the typical effect of temperature and pressure on the fired density.

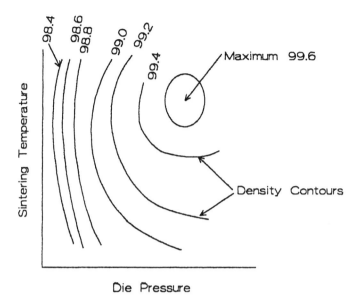

Figure 10.3: Fired density contours as influenced by temperature and molding pressure, TZP zirconia. A maximum density occurs with a drop-off to the left of the diagram.

Figure 10.3 presents more information than first apparent by cursory review. Quite evident is the locus of equal densities depicted by the contours with a maximum in the area shown. The figure is a summary of a statistically- designed, factorial experiment where the interaction is evident from the contours sloping down to the left. An increase in pressure allows the ceramic to sinter to an equivalent density, but at a lower temperature. This is reasonable since the particles are packed closer together. Ordinarily, one maximizes the density. However, other criteria such as fracture toughness, can take precedent.

Somewhat unexpected is that the contours curve around the maximum, where over-firing leads to a reduction of sintered density. This effect can be produced by bloating which will be discussed later. Additionally, the contours are more closely spaced at lower die pressures; that is the slope is steeper. One could conclude that low pressures make pressings where densification is insensitive to temperature. Lacking sufficient green density where the particles are mainly in contact, the material does not shrink as it normally would.

Also from the figure, one can observe that the contours slope around the maximum at the bottom, becoming almost horizontal. Here the sintering almost becomes independent of *pressure*, whereas on the left side of the figure, density was independent of *temperature* (both conclusions limited to this part of the diagram). Why the reversal? It is possible that at higher forming pressures, the particles are so close together that they sinter at a lower temperature resulting in the flat response. Of course, at even lower temperatures they would not sinter at all. However, for this data, the temperature was high enough for sintering to occur.

Density, Die Pressed Alumina

Alumina acts similarly to zirconia. However, when the powder is less reactive, the maximum does not loop around. This effect is potentially due to the higher sintering temperature that desorbs the gases so that bloating does not occur. This is shown in Figure 10.4.

At lower temperatures, the contours level out as observed in the TZP data. The reason for the independence of density and temperature is probably the same, at least for this temperature regime. Again, the contours are more closely packed on the left side of the figure, indicative of a strong dependence on temperature as the green density is increased.

For a given material and with other processing variables constant, contour maps, such as the preceding two figures, allow one to predict sintering behavior.

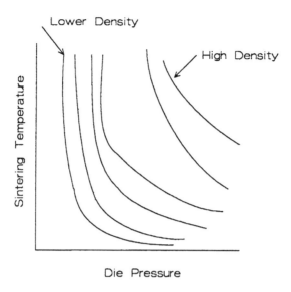

Figure 10.4: Fired Density contours as influenced by temperature and pressure, Alumina. Fine-grained alumina shows the maximum density towards the right of the diagram and, again, sloping off to the right.

Isopressed Alumina

Isopressed alumina of the same type shows a slightly modified behavior, as shown in Figure 10.5.

There is less interaction between temperature and pressure at low pressures. This decreased interaction is possibly due to the uniformly-applied pressure that results in an improved green pack. The contour map is much more symmetrical in form than before. This is probably because the green density is much more uniform due to isopressing that hydrostatically presses the powder. Perhaps this effect can be considered as a fundamental advantage of isopressing over die pressing.

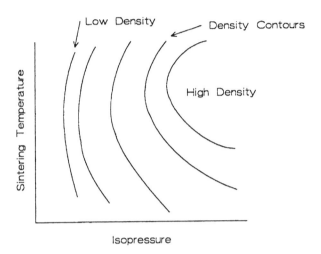

Figure 10.5: Fired density contours as influenced by temperature and pressure, isopressed Alumina. Slope off to the right is not as steep as with die pressed alumina.

Shrinkage, Axial/Radial Die Pressed Alumina

There is little difference between axial and radial shrinkage until the pressure reaches a high value, as seen in Figure 10.6.

As pressure controls the green pack density, it also controls shrinkage when sintering to full density. Of particular interest in the contours is that the axial shrinkage, of say 21%, is achieved at a lower die pressure than that of the radial. Packing is anisotropic with a higher green density along the pressing direction.

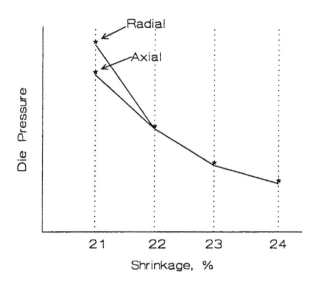

Figure 10.6: Firing shrinkage as influenced by die pressure, Alumina. Axial shrinkage is less, due to better packing at high pressures.

A similar thing happens with coarse-grained ceramics, except that it does not show up as shrinkage. Rather, it manifests as a higher sonic velocity along the pressing direction, due to tighter packing.

There is another conclusion that can be drawn from the data. As pressure increases, so does the spread between axial and radial contours. Since particle packing is tighter along the axial direction, less pressure is needed to obtain an adequate green density for sintering. Friction within the compact retards lateral motion of the particles. This results in green density gradients with a lower overall density that require additional pressure to obtain full shrinkage.

Check List, Effects of T/P

- Combination of T/P: locate maximum density.
- Statistically-designed experiments are preferred. There is an interaction between T/P.
- Too low a P does not sinter to a high density.
- High-forming pressure will sinter to a high density over a range of temperatures.
- Less reactive powders do not densify as readily.
- Isopressed powders are more isotropic in sintering behavior.
- Die-pressed parts shrink more in the diametrical direction than in the axial when the pressure is high enough.
- Packing density in the pressing has a large effect on sintering behavior.

4.0 EFFECT OF TEMPERATURE ON PROPERTIES

Properties that are developed in the ceramic are the result of interactions between the starting powder properties, green structure, temperature, and time. Temperature of itself can affect the grain growth, gas desorption, melting, and phase changes such as those found in zirconia systems.

Gas Desorption

Bonding in many ceramics is very strong. When a crystal fractures, the surface bonds are broken and become extremely reactive. Surfaces are very strong absorbents. High-surface-area powders can absorb significant quantities of gases, which are difficult to remove. These types of powders have gained widespread use, making gas desorption an issue.

Observation of Desorption

When a bisque-fired alumina $(10m^2/g)$ preform is placed in a vacuum furnace and heated, it desorbs gases. Figure 10.7 shows the results of this type of experiment. After pumping the system down, the system is closed and the temperature increased by an increment. The pressure is then measured. This process is repeated over the temperature range of significance. An additional run is made without the ceramic as a blank, and these pressure readings are subtracted from the first.

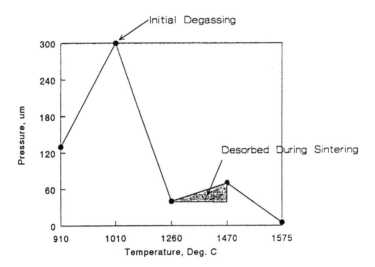

Figure 10.7: Gas Desorption with Temperature, Alumina. Adsorbed gases are not removed below 1260 °C and can be trapped in the structure.

There is a large burst of desorbed gases as the temperature is increased to 1010 °C. Less gas is evolved at 1260 °C, with a second small burst of desorbed gas when heated to 1470 °C. This last burst of gas is shown by the shading in the figure. The gas amount approximately equalled a monolayer desorbed from the ceramic surface. Thermal gravimetric analyses (TGA) have mistakenly fed the notion that gases are driven off at much lower temperatures such as 400 °C.

Bloating

The problem now becomes one of gases entrapped in the structure. This problem results from the inability to desorb gases prior to the ceramic sintering to an impervious state. When such a ceramic is reheated, it bloats. Figure 10.8 demonstrates bloating in alumina that was hot pressed to several temperatures.

Aluminas hot pressed at 1500 °C and 1600 °C both show decreased density when reheated to 1350-1400 °C. Hot-pressing at 1700 °C produces limited initial bloating but levels off above 1300 °C. It appears that the higher hot-pressing temperature allows the absorbed gases to escape by diffusion while the ceramic is still under pressure in the hot press.

It is important to note the following two factors. When the ceramic part is used at an elevated temperature, it will bloat if sintered at a lower initial temperature. It is likely that the grain boundary contains a monolayer of absorbed gas that is most likely water. In the case of alumina, the boundary bonding may be through OH-OH groups, which, if true, would alter the integrity of the structure. These considerations apply only when the sintering temperature is low enough to entrap absorbed gases, which is not always the case.

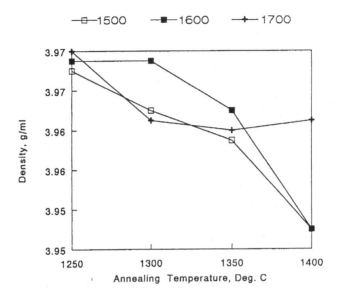

Figure 10.8: Bloating, Hot-Pressed Alumina. Alumina hot pressed to lower temperatures will bloat when reheated due to entrapped gases.

Decomposition

Temperature can also result in the decomposition of some ceramics, especially when one of the molecules has a higher vapor pressure than the other. Three cases are considered below.

B" Alumina. This composition is a sodium-alumina composition that is used in a variety of fuel cells. The problem with its manufacture is that Na_2O evaporates during firing. There are two ways to handle this problem: either pack the ceramic in a bed of the same material or encase the ceramic in a platinum enclosure.[2]

MgO-ZrO$_2$. As previously mentioned, MgO volatilizes from the surface during firing, leaving a monoclinic zirconia layer that is fluorescent under UV light. In one case, the surface of the part was crazed with cracks; this is unusual as a monoclinic surface should result in compressive rather than tensile forces. By sectioning and inspecting with UV, one could explain the crack layer as follows. The layer just below the monoclinic surface was enriched in MgO, making it cubic with a higher thermal expansion than the surface layer. At high temperatures molecular species are mobile, sometimes in unpredictable ways.

SiC/Si$_3$N$_4$. In the case of SiC, the higher vapor pressure atom is Si, which evaporates leaving a graphitic surface. In the case of Si$_3$N$_4$, N$_2$ is the molecule with the higher vapor pressure. In both cases, the ceramic tends to change composition during firing. Again, there are two ways to address this problem: either bury the part in a bed of the same composition or suppress volatilization by increasing the vapor pressure of the more volatile species. For silicon carbide, the furnace interior is presiliconized. In the case of silicon nitride, an over pressure of N$_2$ can be used. When burying the part, it is necessary to use a coarse-grained sand of a material that will sinter to a crumbly consistency. Otherwise, the part will be entombed and difficult to release.

Gas Absorption

The previous discussion showed some of the effects of absorbed gases on microstructure. One must now address the source of these gases.

Absorption from the Atmosphere

One obvious source is the atmosphere. Many a practicing ceramic engineer has personally experienced the whole fabrication process fail in the humid summer months due to absorption of atmospheric moisture. In one case, a calcined powder had to be processed within an hour for it to behave properly.[3] Absorption was inevitable and was only a question of time. All ceramic surfaces absorb water vapor; the quantity of absorption is decreased with large particle sizes but the saturation is the same. This becomes critical when working with ceramics that readily form hydrates, such as MgO and CaO. While a dry box will slow down hydration, it will not prevent hydration from eventually occurring. However, if the processing is done quickly, hydration can become a manageable problem. Often unrecognized are the absorption processes that transpire during firing.

Absorption in a Kiln

Composition of the atmosphere in a gas-fired kiln consists of CO_2, CO, H_2O, O_2, N_2, and possibly some residual hydrocarbons. At elevated temperatures, various molecular species can become volatile and can subsequently be absorbed by the ceramic surfaces. Much of this behavior is predictable from tables on the standard free energy of formation.[4] The following species are suspect: Na_2O, K_2O, B_2O_3, and, in some cases, SiO_2. Water vapor or carbon monoxide can react with these molecules and cause them to become volatile. These reactions are temperature-dependent since the shape of the free energy curves can change with temperature.

As most kilns are lined with refractories containing at least some of these species, contamination inevitably occurs. Unfortunately, these are fluxes that react on the surface or in the microstructure of the grain boundaries as glassy phases. Glass on the grain boundaries decreases resistance to sag at elevated temperatures and decreases resistance to corrosion. Attack along the grain boundaries is not uncommon. Lab furnaces are often lined with high- alumina fiber blocks. These kilns are much cleaner than those lined with fire brick. However, one must proceed with caution. When lab work is done in a clean kiln and production is done in a firebrick kiln, the lab test results may be misleading. These differences can be very large. For example, a ceramic for application in a molten metal had a hundred fold decrease in corrosion rate when fired in a clean environment.

In reducing atmospheres, silica can become volatilized as SiO, which will be absorbed on interior kiln surfaces and on the ware. This is not always harmful as in the case of firing SiC ceramics. As previously mentioned, the graphite must be silicionized to prevent decomposition of the silicon carbide surface. The volatile species reacts with graphite to form a SiC lining.

Gases in Graphite

Any light-colored ceramic will discolor to grey when fired in a graphite-lined kiln. Graphite itself has a very low vapor pressure at the temperature where oxides are usually sintered. For example, at 1627 °C (1900 °K), the vapor pressure of monatomic carbon is 3.47×10^{-12} atmospheres, which is not enough to account for the contamination.[5] However, there are other gases that evolve from graphite when it is heated, including CO_2, CO, and N_2. Some grades of graphite contain as much as 0.115% S. ATJ graphite has only 0.043% S, but the odor is still noticeable. Sulfur absorbed in ceramics is difficult to remove and can have harmful effects on properties such as strength.

Hot pressing in graphite molds places graphite in direct contact with the ceramic, and evolved gases are absorbed onto the ceramic surfaces. An alumina with a surface area of $10m^2/g$, hot-pressed in ATJ at 1700 °C and then abraded by sand blasting exhibited the abraded surface contour seen in Figure 10.9. Please note the 10x vertical exaggeration.

This is a cross-section view, with the graphite contact surface on the left. This edge is abrasion-resistant, possibly because it was cold-worked by shear during hot pressing. The weakened zone extends inward for about 0.1", and then the interior levels off.

There is substantial weakening of the ceramic as indicated by the wear pattern. This is also seen by microhardness indentations where this zone, at its deepest, has twice the incidence of fractures around the indentations than does the interior. The weakening is thus substantial. When the graphite mold cavity is lined with Mo foil, this zone does not appear suggesting that the diffusion is largely retarded. The molecule responsible for this effect has not been identified, but one could hypothesize that it may be sulfur.

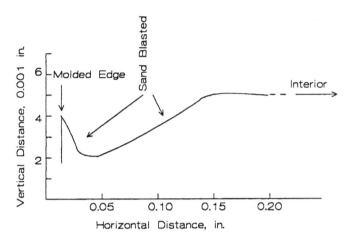

Figure 10.9: Abraded surface, Hot-Pressed Alumina. Sand blasting reveals a weakened zone adjacent to the contact with the graphite.

Mechanical Properties

Strength (MOR) and fracture toughness will be briefly discussed in this section.

Strength

Strength, in particular, is a capricious property that depends upon intergranular phases, internal flaws, residual stresses, and surface finish. These complications make strength difficult to specifically, yet generally, define. Increased temperature results in grain growth, which decreases strength. For example, an alumina over-fired just 100 °C is reduced to half its strength. On the other hand, fracture toughness increases with grain growth. For a TZP ceramic, this increase ranges from 4.5 to 6.0 MPa m$^{0.5}$, when sintered over a range of 250 °C.

Fracture Toughness

Another aspect of the effect of temperature on strength is strength at higher temperatures.[6] Some information is shown in Figure 10.10.

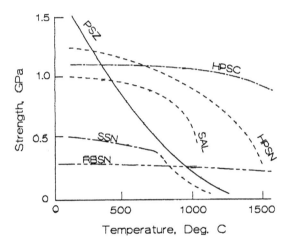

Figure 10.10: High-Temperature Strength, Various Ceramics. HPSC- hot pressed silicon carbide, PSZ- MgO partially stabilized zirconia, HPSN- hot pressed silicon nitride, SAL- sintered alumina, SSN- sintered silicon nitride, RBSN- reaction bonded silicon nitride.

Most of this data is for materials being investigated for high-temperature use such as in ceramic gas turbines. The curves have been redrawn and do not include all the data from the reference since this figure is intended only to illustrate general behavior. In general, strength drops off with increasing temperature. Hot-pressed silicon carbide is exceptional in its resistance to temperature.

Specific data is not especially helpful since properties change depending on materials and processing conditions. The authors of the reference indicate that high-temperature strength depends largely upon the nature of the boundaries, with grain boundary glass having a negative effect. Oxides such as alumina decrease in strength where diffusion rates start to become appreciable around 1200 °C. Here, too, glass on the grain boundaries decreases strength at elevated temperatures.

Effect of Temperature on Grain Size

It is commonly known that an increase in temperature results in grain growth. This grain growth is beneficial if the desired effect is a decrease in transmitted light scattering. However, grain growth is undesirable for conditions of higher wear resistance. A very important consideration in many ceramics is maintaining pores on the grain boundaries where they can be sintered out. Pores within grains are largely trapped. Control of pore location relates to the rate of grain boundary movement. At least four factors are important: the starting powder, green structure, heating rate, and grain growth inhibitors.

With alumina, MgO is added to control grain growth. When the starting powder is right (i.e., correct sub-micron particle size, high purity, high uniform green density, and deagglomeration), the heating rate is not as critical, but restraints still exist. Figure 10.11 illustrates the inception of uncontrolled grain growth in alumina without MgO.

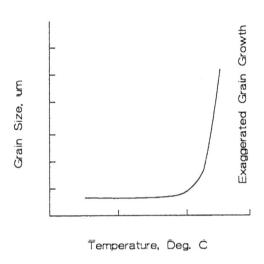

Temperature, Deg. C

Figure 10.11: Grain Growth with Temperature, Alumina. Exaggerated grain growth occurs when the temperature reaches a threshold value.

The curve is flat until about 1600 °C where grain growth significantly increases. The examined microstructure reveals that some grains grow at the expense of others and often become faceted. While it is possible that this might increase fracture toughness, it is deleterious to other properties such as strength. To avoid discontinuous grain growth, one should examine the following: the starting powder properties, limited sintering temperature, grain growth inhibitors such as MgO or NiO for alumina, or included second phases that drag on the boundaries as they move.

Check List, Effects of Temperature on Properties

- Gas Desorption
    High temperature needed
    Bloating
    Surface decomposition

- Gas Absorption
    From the atmosphere
    From the kiln
    From graphite
- Effects on Mechanical Properties
    Strength
    High temperature strength
    Fracture toughness
- Effect on Grain Size
    Grain growth inhibitors

5.0 EFFECTS OF PRESSURE ON PROPERTIES

Pressure used to compact a green body directly affects green density which, in turn, influences firing shrinkage, specific gravity, and wear resistance.

Green Density

Green density increases as the die pressure increases, as shown in Figure 10.12.

The curve is typical for a fine powder, with green density increasing with pressure and then starting to level off at high pressures. With a rigid test machine, the curve will show one or two breaks. This is actually what occurs but is not observable with an ordinary press. When the plot is made with percent of theoretical density plotted against the logarithm of pressure, the curves become straight lines, as shown in Figure 10.13.

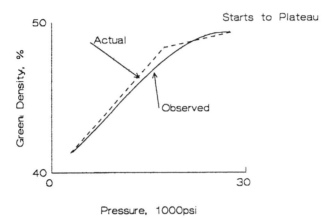

Figure 10.12: Green density with Die Pressure. Green density plateaus as pressure is increased.

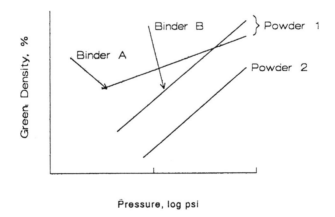

Figure 10.13: Green Density with Log Die Pressure. Powder 1 compacts more readily than powder 2. The two binders act differently with pressure.

Different binders or powders exhibit different slopes and displacements in the figure. In this illustration, the #1 powder compacts to a higher density than #2. Highly sinterable powders sinter to a high-fired density even at relatively low green density (45%). Less sinterable powders are more sensitive to the green density and usually require a higher green density to sinter properly. The figure shows that selection of the powder and binder are tools for improving compaction.

Compaction

The powder compacts as it is pressed. The ratio of the fill height to the compact height is the *compaction ratio*. For a fine powder, this ratio is around 0.4. As the part is ejected from the mold, it *springs back* to a larger diameter than that of the die cavity. For a fine powder, the spring back is about 0.4% depending on how hard it is pressed.

Shrinkage

Firing shrinkage varies with the molding pressure. It is intuitive that the more dense the pressing, the less shrinkage occurs. A typical curve is shown in Figure 10.14.

This curve is for a fine compact material. Coarse materials have little or no firing shrinkage. Packing is more dense in the pressing direction, so there is less shrinkage. As a general rule, higher shrinkages result in increased firing distortion, less size control, and more cracks. On the other hand, high molding pressures can result in green cracks from a variety of sources. For any system, the optimum molding pressure must be worked out.

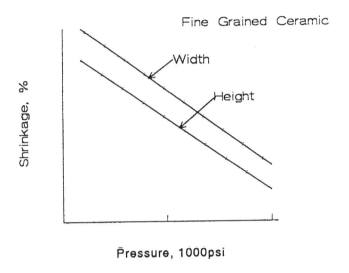

Figure 10.14: Firing shrinkage with Die Pressure. The width of a cylindrical part shrinks more because it is packed to a lower green density.

Fired Density

For a given firing temperature, the sintered density is a consequence of the molding pressure as shown in Figure 10.15.

This figure suggests that the particles in the compact must be close enough together to sinter well. When the molding pressure is too low, the compact does not sinter to full density, and the pore size is relatively large due to pore coalescence. The curve begins to level out at higher pressures as the particles impinge upon one another.

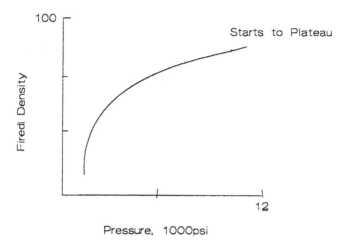

Figure 10.15: Sintered Density with Die Pressure. Fired density increases with die pressure and then plateaus.

Wear

With pressure as the variable, wear resistance is affected by the fired density. High-density ceramics wear less than the same material at a lower density, as shown in Figure 10.16.

The figure shows the cumulative percentage weight loss plotted against time of the wear test. The two curves bracket an under-pressed body and a properly-pressed body. If additional pressure is used, the wear resistance levels out since adequate green density has already been achieved.

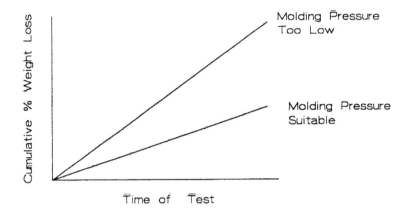

Figure 10.16: Wear resistance with Die Pressure. Weight loss is greater when the part is molded to a lower pressure.

Check List, Effects of Pressure

- Green Density
    - Binders
    - Powders
- Compaction
    - Compaction ratio
    - Spring back
- Shrinkage
- Specific Gravity

6.0 EFFECTS OF MICROSTRUCTURE ON PROPERTIES

Ceramics, as a technology rather than a science, are highly unpredictable. In fact, the generalization that, "Every case is special," often holds true in the realm of ceramics. This section presents some examples to illustrate this often-capricious behavior. Since these examples are not intended to be read quantitatively, only some values are given in the graphs to bring the illustration to scale.

Porosity/Properties

Strength

As expected, strength decreases as the volume fraction of pores increases.[7] This is shown in Figure 10.17.

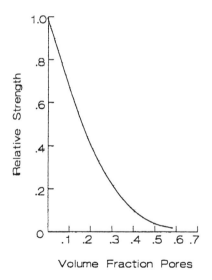

Figure 10.17: Strength with Volume Fraction Porosity. Strength drops rapidly as the porosity increases.

The curve applies to a variety of materials. For most ceramics, the part will not hold together when the volume fraction of pores exceeds about 40-50% for a normal microstructure.

Elastic modulus

The elastic modulus follows a similar relationship as strength, with the modulus decreasing as the pore volume increases.[8] This is shown in Figure 10.18.

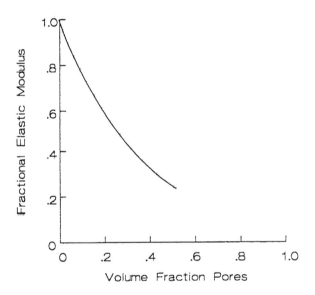

Figure 10.18: MOE with Volume Fraction Porosity. The modulus of elasticity also drops off rapidly as porosity increases.

Thermal shock resistance increases with an increase in percentage pore volume due to the lower modulus.

Grain Size/Properties

Grain size (measured as the intercept on a polished surface) affects the physical properties of strength, fracture toughness, and hardness. The information shown in the next two figures is from an alumina ceramic made from a sub-micron powder which was sinterable at fairly low temperatures.

Strength

Over-firing a ceramic substantially decreases the MOR. A 100 °C increase in the soak temperature decreases the strength from 632 MPa to 337 MPa while increasing the grain intercept from 1.4 μm to 4.8 μm.

Fracture toughness

Figure 10.19 provides statistics on both the grain intercept measurements and fracture toughness. Regarding the measurements, CV is the coefficient of variation and SD is the standard deviation. Standard deviation increases as grain size increases. While this is expected since the measurements have larger dimensions, it is somewhat misleading. Dividing the standard deviation by the mean grain intercept normalizes the results such that there is even a slight decrease at the coarsest size. This result is consistent with observation, as the smallest grains have disappeared in the microstructure. Fracture toughness increases as the grains grow larger. Even though the increase is small, it is significant.

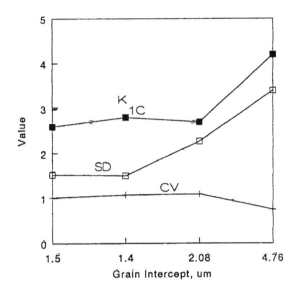

Figure 10.19: Fracture Toughness with Grain Intercept. K_{IC}-fracture toughness, SD- standard deviation, CV- coefficient of variation.

Hardness

Figure 10.20 shows the relation between Vickers hardness and grain intercept.

Hardness is flat until the microstructure begins to show exaggerated grain growth. The drop is fairly large. Figure 10.21 depicts the relationship between firing temperature and grain intercept.

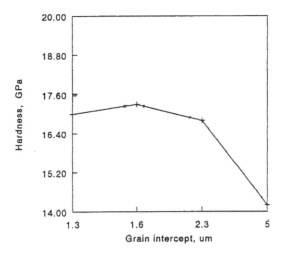

Figure 10.20: Vickers Hardness with Grain Intercept. Hardness drops off as the grain size increases.

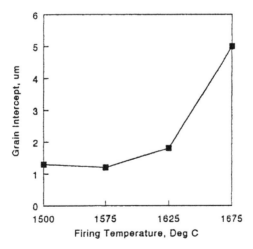

Figure 10.21: Effect of Temperature on Grain Intercept. Grains in this material start to grow rapidly above 1625 °C.

Grain growth is flat up to 1550 °C and then starts to increase. The temperature rise of 50 °C to 1650 °C results in substantial growth. For most applications, this ceramic is over-fired. The microstructures of the 1550 and 1650 °C materials are shown in Figure 10.22.

Note that there is a five fold increase in magnification between the two photographs in Figure 10.22. Both photos are essentially fully dense. Measurements on grain intercept were made on these photos and two others at the other sintering temperature. Evidence of exaggerated grain growth is seen in the 1650 °C firing temperature photo.

Wear Resistance

Wear of TZP using a pin and disc configuration is shown in Figure 10.23.[9]

Wear is strongly influenced by fracture toughness, showing as great as a fourth power dependence as discovered by the authors. Tough TZP is quite wear resistant, especially when sliding against itself. This is a special case where wear rate correlates with a physical property. There exist additional cases where chemical reactions between two different materials control the wear mechanism. Because wear is a rather capricious property, there is no good substitute for a test.

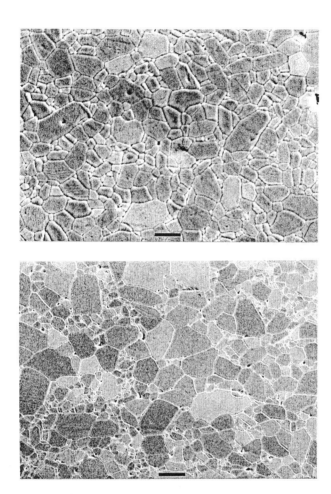

Figure 10.22: Microstructure Alumina Fired at 1550-1650 °C. Note the change in magnification between the two figures. Scale bar 10 μm on both.

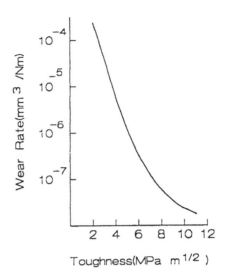

Figure 10.23: Wear Rate with Fracture Toughness, TZP. Wear decreases very rapidly as the fracture toughness is increased.

Phase Composition

MgO-stabilized zirconia is used as an example in the discussions below. Consider the phase diagram of the ZrO_2-MgO system.[10] This is shown in Figure 10.24.

For a structural ceramic, sintering is in the cubic region of the phase diagram, close to 1800 °C and resulting in a 40 μm grain intercept. During aging in the tetragonal/cubic region, the amount of tetragonal phase varies and depends on the amount of MgO. These conditions have a pronounced effect on microstructure properties. Figures 10.23 through 10.26 are samples from the same experiment and can be compared with each other.

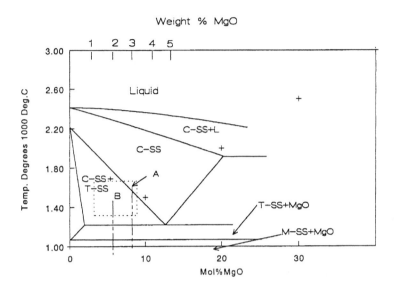

Figure 10.24: ZrO$_2$-MgO Phase Diagram. C-SS cubic solid solution, T-SS tetragonal solid solution, M-SS monoclinic solid solution.

MgO %/Strength

Figure 10.25 shows the strength (MOR) behavior and thermal expansion to 1300 °C with varying MgO content for a fully dense structural ceramic with a 40 μm grain intercept.

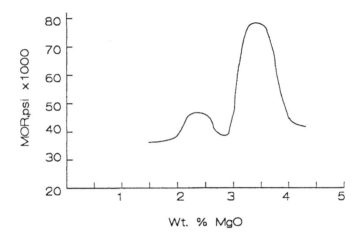

Figure 10.25: MOR with MgO Concentration, ZrO_2-MgO system. The strength has two maxima.

The curve has two maxima at 2.6 and 3.5 w/o MgO. The first maximum at 2.6 w/o MgO is due to the elimination of free-standing, tetragonal grains that transform to monoclinic at lower temperatures. The volume change during this transformation weakens the structure. Further increasing the MgO content shifts the composition to the right. Optimal strength occurs at 3.5% MgO due to the best ratio of tetragonal inclusions in the matrix of the cubic grains. Further additions of MgO decrease the tetragonal content such that toughening is eliminated and the ceramic becomes embrittled. The 3.5% MgO maximum has a steep left side; another species competing for MgO (such as SiO_2) shifts the composition off the crest and to the left. Due to this sensitivity, small amounts of impurities greatly affect MOR. Because strength derives greatly from composition, manufacture must be carefully monitored and controlled to obtain the highest strength. The 80,000 psi maximum shown is for a sample under optimal conditions.

Monoclinic%/MOE/Thermal Expansion

By continuing to decrease the amount of MgO added, the monoclinic content increases as seen in Figure 10.26.

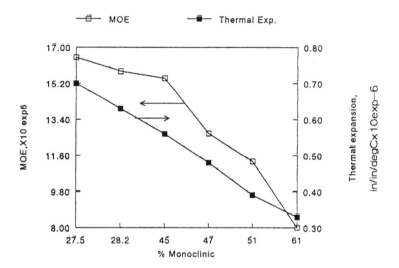

Figure 10.26: MOE/Thermal Expansion, with Percent Monoclinic ZrO_2. Both the MOE and thermal expansion decrease with increasing monoclinic phase.

As shown in the figure, both the modulus of elasticity and thermal expansion greatly decrease as the monoclinic content increases. As both these parameters decrease, the resistance to thermal shock increases. MgO-stabilized zirconia ceramics towards the right side of the composition are used in foundries where molten metal is poured over the cold lip of the crucible. Using such a technique, the crucible withstands a number of cycles before breaking.

Monoclinic%/MOR

This data is taken from a less dense body than that of Figure 10.23, and the monoclinic content is extended to much larger amounts. Figure 10.27 shows the MOR of the as-fired body and that after cycling between 260 °C and 1370 °C.

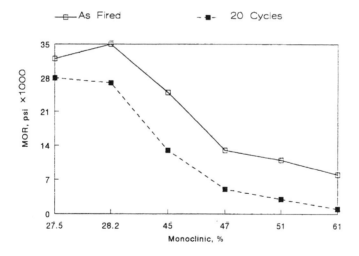

Figure 10.27: MOR-Thermal Cycling with Percent Monoclinic ZrO_2. Strength decreases as the amount of the monoclinic phase increases.

The two curves are parallel, with the MOR decreasing substantially with 20 cycles. Cycling over the transformation temperature loosens intergranular bonding due to the volume change in the monoclinic/tetragonal phase. The foundry crucibles discussed in the last section eventually fail due to decreased strength.

Monoclinic%/Porosity-Cycles/Expansion

Figure 10.28 provides further evidence of the body's disintegration after cycling.

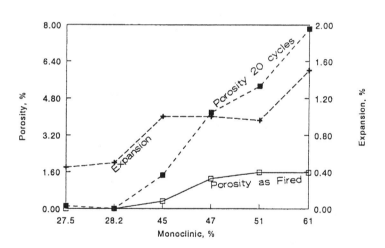

Figure 10.28: Expansion/Porosity-Cycling with Percent Monoclinic ZrO_2. Porosity increases with thermal cycling as the amount of the monoclinic phase increases.

The body does not sinter as easily with a two-phase composition, and porosity results. This effect is visible in the figure where porosity increases from 0% to 0.4% at 61% monoclinic. The results are dramatic after the ceramic is thermally cycled, increasing from 0 to 8% after 20 cycles. There is an attendant expansion of the part's dimensions after cycling, from about 0.4% to about 1.6%. This provides evidence of the body coming apart due to the repeated transformation.

Check List, Microstructure/Properties

- Almost every case is special as there are so many variables.
- Porosity decreases strength, becoming very small at about 40v/o porosity.
- Porosity decreases the elastic modulus that becomes difficult to measure at about 50v/o porosity.
- Fracture toughness increases somewhat as the grain size becomes larger. The coefficient of variation is essentially flat.
- Hardness is flat as the grain size increases up to a point, and then drops significantly.
- Grain size (intercept) is flat as the sintering temperature increases up to a point and then increases steeply.
- The phase diagram of the ZrO_2-MgO system charts the predicted behavior of the ceramic.
- MOR in the ZrO_2-MgO system shows two maxima, with the larger one at about 3.5% MgO.
- MOR in this system is very sensitive to the amount of MgO available for stabilization.
- Properties in the ZrO_2-MgO system are very dependant on the amount of the monoclinic phase in the ceramic. MOE drops linearly from 27.5% M to essentially 0 at 61% M. Thermal expansion drops linearly with monoclinic content, the curve at high monoclinic content has two humps that lower the overall expansion.

- Thermal cycling of a 61% monoclinic body over the transformation temperature lowers the MOR to almost 0 after 20 cycles.
- Thermal cycling increases the porosity in this system at high monoclinic contents.
- The dimensions of a high monoclinic content body become greater after cycling.
- The ZrO_2-MgO systems properties are sensitive to impurities that compete for MgO.

REFERENCES

1.   Fred F. Lange, "Powder Processing Science and Technology for Increased Reliability," *J. Amer. Ceram. Soc.* 72[1]3-15 (1989).
2.   Ronald S. Gordon, Personal Communication.
3.   Morris Berg, Personal Communication.
4.   O. Kubaschewsky and E.L. Evans, Metallurgical Thermochemistry, Pergamon Press.
5.   The Industrial Graphite Engineering Handbook, Union Carbide Corporation.
6.   Yohtaro Matsuo and Shiushichi Kimura, "Statistical Evaluation and Strength Data of Engineering Ceramics," *Jour. Ceram. Soc. Japan*, **95** pp C400-409.
7.   W. D. Kingery, Introduction to Ceramics, John Wiley & Sons, p622.
8.   Ibid, p598.
9.   Soo W. Lee, Stephen M. Hsu, and Ming C. Shen, "Ceramic Wear maps: Zirconia." *J. Amer. Ceram. Soc.* 76[8] pp1937-47 (1993).
10.   V.S. Stubican and J.R. Hellmann, "Phase Equilibria in Some Zirconia Systems," Science and Technology of Zirconia, Edited by A.H. Heuer and L.W. Hobbs (Amer. Ceram. Soc. 1981).

11

Ceramic Property
Measurements

1.0 INTRODUCTION

This chapter covers analytical, slip properties, microscopy, and physical properties with emphasis placed on using these methods to characterize materials and densified ceramic articles. Properties provide clues to the history of the ceramic including how it was made, processes that occurred during its manufacture, and its environmental interactions during use. Also discussed in this chapter are the crafts used in these methods and in sample preparation.

2.0 ANALYTICAL

This section describes common analytical methods used to characterize ceramic materials. These methods include chemical composition, structure, surface area, thermal, particle size, and surface analytical methods. Since a comprehensive description of these methods is beyond the scope of this chapter, this discussion instead focuses on the function and application of the methods, craftsmanship in preparing samples, and making measurements.

Analytical measurements pose the serious problem of obtaining standards. Ideally, standards should derive from the material being analyzed.

For example, when measuring silica in a ZrO_2 powder, the standards should have chemistry over the same range and have the same phase composition as the sample. This is easily accomplished when the standard can be batched, which is sometimes the only option available. A hazard that must often be accepted is that anything done to the sample can change its chemistry.

Spectrographic

The spectrographic methods used here describe instances where some kind of spectrum is utilized for the measurement. Emission spectra are more widely used for analysis of ceramics than absorption spectra. While atomic absorption was an important absorption method for chemical analyses, it has largely been superseded by induction-coupled plasma (ICP), an emission technique. Other absorption spectra methods such as infrared are not commonly used with ceramics.

Arc Emission

The sample can be a wire, chunk, shaving, or can be pulverized, often in a capsule containing a metal ball. Contamination from the capsule and ball occurs as the assembly vigorously shakes in a machine. To circumvent this problem, the capsule and ball composition should differ from that of the sample materials analyzed. This is easily done for ceramics that are mostly comprised of light elements while the capsule is often WC/Co or steel. While a plastic capsule can be used, such a ball is not massive enough to crush most ceramics. A small B_4C mortar and pestle can often be used. Larger pieces can be broken up in a Plattner mortar and pestle that is impacted with a hammer. However, pieces of the ceramic become embedded in the base and cause contamination in the next sample. This contamination can be reduced by first crushing a piece of the sample and then throwing it away. The next piece will have less contamination when crushed. Safety concerns also exist as there is the potential to break one's finger or thumb with the hammer should one's attention be momentarily diverted. While there are many other lab-sized crushers available, they are designed for materials much softer than most ceramics. Furthermore, these crushers all contaminate the sample.

With the sample now as a powder or lump, it can be packed or placed in a recessed graphite electrode and placed in an arc chamber. The arc vaporizes the sample at a temperature high enough to ionize the elements. Each element has a unique spectra by which it can be distinguished from most others. Because some interferences exist, not all elements can be analyzed by this method. The spectrographer simply indicates what can and cannot be done. Light from the arc is broken up into spectra with an Echelle optical design and read with a charge injection device (CID). The latter device has superseded photographic plates. An Echelle optical system consists of a shutter, mirror, prism, the Echelle grating, a mirror, and the CID detector. Since the light path folds back on itself, the instrument is smaller than the older spectrographs and can be bench-mounted. The spectrographer's care largely governs the method's accuracy. Accuracy is reported in ranges from major/minor/trace, to decades such as 1%-0.1%, to single digit values such as 2%. While single digit numbers often suffice, most spectrographic labs lack adequate standards for the specific material being analyzed and thus cannot achieve these numbers. Though standards are most important, other factors such as vibration, lab temperature, stability of the arc, and spattering of the melted sample also contribute. Here, meticulousness in technique is an asset.

An additional advantage of the arc emission spectrograph is that one can determine the rough composition of a very small sample, thereby following the motto that, "If it can be seen, it can be analyzed." An SEM with analytical capability proves even more useful. An emission spectrograph is shown in Figure 11.1.

Induction Coupled Plasma (ICP)

In this method, plasma is used to dissolve and vaporize the sample. The plasma is heated to a very high temperature with an induction coil. Since the coil is not in the plasma, it is non-contaminating. The high temperatures ionize the elements dissolved in the solution. This technique produces an emission spectra that can be dispersed with Echelle optics and sensed with a CID, similar to the arc emission spectrograph previously described. An ICP instrument is shown in Figure 11.2.

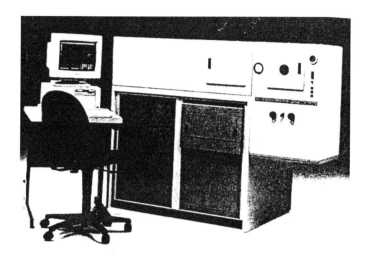

Figure 11.1: Emission Spectrograph. Generally useful for analyzing many elements down to trace quantities. (Courtesy of Baird)

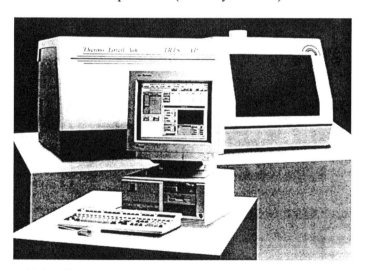

Figure 11.2: ICP Instrument. Widely used for chemical analysis. (Courtesy of Thermo Jarrell Ash)

When trace elements are important, the chemical purity of the flux is of concern as it empties into the plasma along with the sample. Many ceramics are fused in a flux since they are insoluble in acid. When this interference exists, the flux impurity concentration must be analyzed in the same way and then subtracted from the sample results. Although this method decreases the precision of the analysis, it is the best option.

It is problematic that in dissolution of ZrO_2 for a silica determination, silica can escape as SiF_4, which has a high vapor pressure. One technique involves dissolving the zirconia in a Teflon-lined bomb with HF at an elevated temperature and subsequently chilling the bomb in ice water before it is opened.

X-Ray Fluorescent Analysis

The sample is either a compacted powder or a solid, flat specimen irradiated with X-rays. In the example, the tube has a Rh target, but other targets are available. Atoms in the sample become excited and fluoresce with their individual spectra that is analyzed with a diffractometer. The intensity of the spectral lines is proportional to the amount of the element in the sample. This intensity is reduced by adsorption of other elements in the sample, rendering a lower concentration reading than actually exists. As with all analytical techniques, interferences in the spectra do exist. Despite these interferences and adsorption errors, this method is otherwise non-problematic and is generally useful. That the analyst is aware of the presence of both adsorption and interference makes these errors predictable, which is preferable to other methods where potential problems remain unknown.

Calibration standards are not as problematic because the amount of the element, rather than the phase, comes into play. Since weight analysis is very precise, the sole problem is to uniformly distribute the element in the sample, which becomes a mixing problem. To test the mixing technique, mix a small amount of red iron oxide pigment into a white powder of similar particle size. Take a small amount of the mixture and smear it onto a piece of white paper with your finger. Red streaks indicate that the pigment is still agglomerated and that it requires more shear to mix the material uniformly.

One good technique is to use a mortar and pestle, with frequent scrapping down of the material on the sides. Wetting the sample with an organic solvent while grinding helps to obtain uniformity. Acetone is useful, but be sure to continue mixing until the sample is at least dried to a paste, or it can segregate.

A mandatory practice is to make a concentrate batch when mixing a small amount of one material into another. For example, when making a 0.1% standard of an element, start by making a 10% master batch. After this is thoroughly blended, dilute it to a 1.0% master batch that can in turn be used to make the final composition. It is much easier to mix two materials uniformly if they are of reasonably equal volume. A 10:1 mix is not too bad but 1000:1 is not acceptable.

Compensate for adsorption by using an internal standard. Mix in a known amount of another specific element with a spectral line close to the one being measured. One can now make a correction because the line will be adsorbed the same amount. In addition, one can use four different diffracting crystals, which can be built into the apparatus, to cover a wide band of wave lengths. This technique has its time drawbacks since spectrum scan is slow. When interested in just one element, set the diffractometer on that angle for speedy data gathering. By doing so, analyses are nearly made in real time, and the technique is useful as an in-line process control.

X-ray fluorescence can be used to analyze trace amounts of elements. To do this properly, set the spectrometer angle just to one side of the peak to be analyzed and take a count of the background radiation intensity. A long count provides precision. Accuracy is further improved by measuring the background at the other side of the peak and averaging. Then, set the spectrometer on the peak of the line of the element to be measured and take a count. The first time this is done, find the exact angle of the peak by running a traverse across the peak to find the top, set the angle on the top of the peak, then make a count. Now, subtract the background count from the peak count and compare with the standard for the analysis. The combination of a properly mixed internal standard, correct location of the peak maximum, and a long count of radiation intensity makes X-ray fluorescent analysis the most accurate analytical method for general use. Figure 11.3 is a drawing of a X-ray spectrometer.

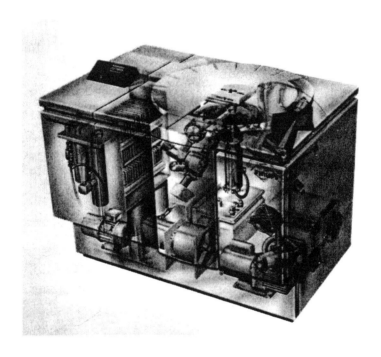

Figure 11.3: X-Ray Spectrometer. Can be very accurate when using standards. (Courtesy of Philips)

Several key components are numbered in the drawing.

#1. The X-ray tube has a ceramic body instead of glass and is configured so that the Rh target is close to the sample, resulting in increased sensitivity.

#2. An optical encoder is used to precisely measure and control the position of the goniometer.

#3. Two types of X-ray detectors can be used, and are placed close to the diffracting crystal for improved sensitivity.

#4. There are two fixed channels that help to reduce measurement time.

#5. The X-ray beam is collimated with three choices for sensitivity vs. resolution.

#6. The primary beam is filtered to increase the sharpness of the spectra.

#7. The sample is surface down, allowing analysis of liquids.

#8. Multi-channel electronics

#9. Precise instrument temperature control

There is reason for the lengthy description provided above. While modern instruments are sophisticated and improved versions of older models, one must nonetheless proceed with care since the "bells and whistles" that reduce labor and analysis time do not always add to the raw data in a meaningful way. In the example, all nine features either improve the crispness of the data or aid in its presentation.

X-Ray Diffraction

X-ray diffraction produces an array of peaks resulting from lattice geometry rather than a spectra of wavelengths. This technique is used to identify the crystalline phase present in the sample. Each phase has a nearly unique diffraction pattern; when the same pattern appears for more than one phase, chemical analysis can aid in sorting. Diffraction is widely used for crystal structure analysis, but is not commonly used in the ceramic laboratory.

X-ray analyses are used for phase identification, particle or grain size analyses, presence of glass, colloidal particle size, and stress analysis.

Powders are analyzed for the phases present by a diffractometer. The apparatus is shown in Figure 11.4.

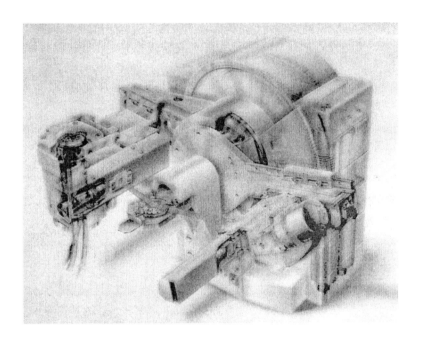

Figure 11.4:   X-Ray Diffraction Apparatus. Used for phase analysis. (Courtesy of Philips)

X-rays from a source pass through a monochrometer, collimated, and impinge upon the sample being analyzed. The sample is usually powder-packed in a rectangular holder, but can also exist as a film on a glass slide or a solid ceramic planer surface. Interplaner spacings in the crystal lattice diffract the beam at different angles, producing a spectrum. A sketch of the geometry is presented in Figure 11.5.

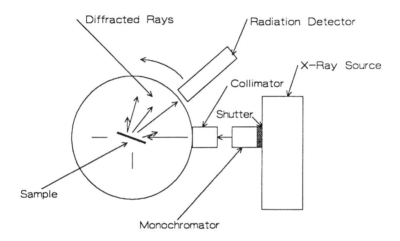

Figure 11.5: Sketch, X-Ray Diffraction Geometry. Both the detector and sample rotate during the measurement.

The angles in the instrument shown in Figure 11.4 are measured with a shaft-encoder-controlled goniometer that contains a radiation detector to measure ray intensity. Slit widths are computer-controlled. The amount of a phase present is proportional to the intensity of the diffraction line. One must assume that the particles in the sample are randomly oriented: a poor assumption for elongated or planer particles. In such cases, the phase can still be identified, but the amount can only be estimated. It is difficult to prevent particle orientation when the geometry is unfavorable. One solution is to dust the powder onto a sticky-tape surface to which the particles adhere: an improved yet imperfect solution.

Debye-Scherrer. This older method employs a photographic technique. The sample is a thin rod of powder in a glass capillary or rolled into a spindle with glue. The photographic film surrounds the sample in a coaxial cylindrical configuration. Diffraction by the lattice produces a set of lines on the film that can be measured at each diffraction angle. It is advantageous that a single particle can be used with this technique, following the motto that "If it is big enough to see, it is big enough to X-ray." This is a useful micro technique. Figure 11.6 is a sketch of the geometry.

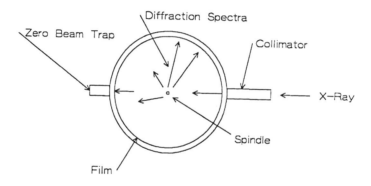

Figure 11.6: Debye-Sherrer Geometry. Photographic technique useful for small samples.

The speck is glued with a cellulose acetate cement (Duco) to the end of a glass fiber or glue spindle. A glue spindle is rolled between two glass slides just when it has dried enough to retain plasticity without being fluid. The spindle must be amorphous so that it does not diffract its own pattern.

Glasses. Since glasses are amorphous, they do not produce a diffraction pattern. However, they do produce a broad hump in the spectrum at the lower part of the diffraction pattern. The amount of glass can be estimated from the size of the hump using standards.

Particle Size. With small particles, measurements of line broadening can be used to estimate size. While electron microscopy is a preferred technique, there may be instances where X-ray diffraction is also useful. However, this method does not provide information on size distribution.

Stresses. When the material is under stress, the crystal lattice distorts, displacing the position of the diffraction lines. The amount of displacement relates to the stress. Internal stresses arise from a number of causes: thermal expansion anisotropy within a grain, expansion coefficients between different materials, compositional gradients, and mechanical work such as grinding the surface of a ceramic article. Residual stresses can be as high as 30,000 psi in ground surfaces.

Check List, Spectrographic Analyses

- Proper Standards
- Arc Emission
    - Many elements
    - Grinding contamination
    - Small sample
    - One significant figure at best
- ICP
    - Many elements can be analyzed
    - Have to dissolve sample
    - Flux purity
- X-ray Fluorescent Analysis
    - Powders, liquids, or flat plate
    - Mixing standards

Internal standards
Precise peak count procedure
Modern instrument design
- X-ray Diffraction
Phase analysis
Powders or flat plate
Small sample possible
Grain size
Glass
Stresses
Modern instrument design

Wet Chemical Analysis

The wet chemical analytical methods are the underpinning of analytical procedures. They are still very useful when the analyst knows what he/she is doing. Procedures are often described in great detail because much can go wrong. Material manipulation always hazards loss or inadvertent addition of other material, such as a co-precipitation in a gravimetric analysis. With any analytical procedure, it is wise to submit blind samples that have been previously analyzed to see if they can be replicated. One can then calculate an error variance. A sharp analytical lab will track their accuracy this way so as to discuss how the methods can be improved. While ideal, checks on accuracy are often construed as professional insult. In this case, the lab is mismanaged and one should find a better source for data if possible.

Another key feature of a keen lab is the involvement in round robins with other labs through which identical samples are analyzed and results shared. Analytical chemistry is part art, part technology, part craft, and has earned much respect.

Wet chemical analyses encompass a variety of methods, including gravimetric, colorimetric, volumetric, and titration. The analytical lab usualy decides which method to use unless the type of sample is beyond their realm of experience.

Check list, Wet Chemical

- Material has to be dissolved.
- Submit blind samples.
- Keep a running audit on accuracy.
- Have the lab participate in round robins.
- Chemist will select the best method.

Surface Area

The BET method of measuring surface area is pervasive. With this method, the sample is baked out in a vacuum to remove absorbed gases such as water vapor. The surfaces are then saturated with a monolayer of N_2. The amount of nitrogen needed to accomplish this is a measure of the surface area. The longest part of the analysis is the bakeout procedure. Equipment is available with ganged bakeout stations to facilitate this. When the surface area is very low, krypton is used instead of nitrogen. Having a higher density, the sensitivity of the analysis is increased. For very fine powders (d = 0.2 μm), surface area is a more sensitive measure of powder activity than particle size. BET data is often used as a process control in these cases. Figure 11.7 is of a BET apparatus.

Check List, Surface Area

- The BET method is used.
- Nitrogen absorption is common.
- Krypton can be used for very fine materials.
- When there are a lot of samples, ganged bake out is useful.
- Surface area is a more sensitive indicator of powder activity than particle size when the size is very fine.

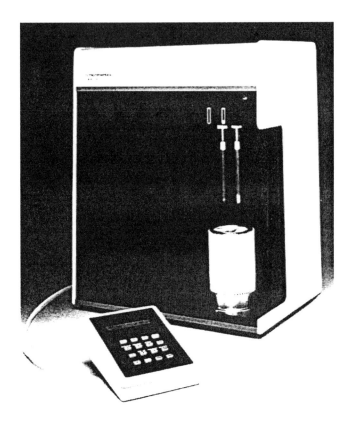

Figure 11.7: BET Apparatus. Measures surface area. (Courtesy of Micromeritics)

Thermal Analyses

Thermal properties of ceramics and refractories are very important for many applications. Thermal analyses include: thermal gravimetric (TGA), differential thermal (DTA), differential scanning calorimetry (DSC) that is similar to DTA,   thermal conductivity, specific heat, thermal expansion, and loss on ignition (LOI).

Thermal Gravimetric (TGA)/Differential Thermal (DTA)

TGA measures the change in weight as the sample is heated in an atmosphere of choice. Equipment is capable of reaching above 1500 °C and usually has an air, nitrogen, or an argon atmosphere. TGA is useful in studies such as binder burnout, desorption, dehydration, and thermal decomposition. DTA measures the temperature difference between a sample and an inert material as they are heated. This is useful for locating the temperature where reactions occur such as for melting, phase changes, and oxidation. Instruments can measure both the change in weight and the change in temperature. Figure 11.8 shows one kind of TGA/DTA apparatus.

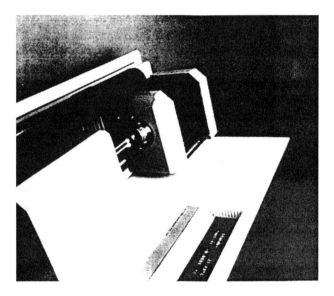

Figure 11.8: TGA/DTA Apparatus. TGA measures weight loss on heating. DTA measures temperature changes relative to an inert standard, as the sample is heated. (Courtesy of TA Instruments)

Thermal conductivity

Measuring thermal conductivity requires both thermal diffusivity and specific heat. Laser pulse method easily measures diffusivity while calorimetry determines specific heat. An instrument used to measure thermal conductivity by the laser pulse method is shown in Figure 11.9.

Figure 11.9:   Thermal Conductivity Instrument. Laser flash measures thermal diffusivity. (Courtesy of Holometrics)

The measurement is made by pulsing a known amount of energy from a laser onto the sample and then measuring the temperature rise. There is a complicated ASTM method for refractories. Barring the need to make many thermal conductivity measurements, it is prudent to farm the work out to a commercial laboratory. Because anisotropic materials often align in the microstructure, they result in different direction-dependent thermal conductivities. For example, thermal conductivity in graphite is much greater across the platelet than along the c-axis. When the platelets are aligned parallel to the refractory wall, the refractory is insulating. When the platelets are aligned perpendicular to the refractory wall, the refractory is thermally conductive. Either can be done, depending on the forming method.

Sometimes, thermal conductivity determines the material of choice for a given application. For example, MgO has a high thermal conductivity and is used as electrical insulation in Calrod heaters. Zirconia has a low thermal conductivity and is used for thermal insulation at high temperatures. Also, materials with a high conductivity are better in thermal shock than insulators.

Porosity has a pronounced effect on conductivity, which is one reason why fibers are used for kiln insulation. Closed, small bubble foams are very good insulators, but are not structurally stable at high temperatures since they shrink and warp. Figure 11.10 presents some data sketched from the reference for alumina in dense, foamed, and fibrous forms.[1]

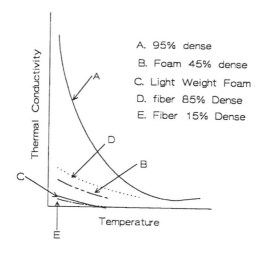

Figure 11.10: Thermal Conductivity Alumina. Thermal conductivity is highly dependent on form of the structure.

Specific Heat

Generally, specific heat is measured when thermal conductivity is determined. In applications, specific heat contributes towards how quickly a material warms and cools. Specific heat is measured with a calorimeter.

Thermal Expansion

Thermal expansion is measured with a dilatometer where a sample of determined length (commonly 1") is heated and the change in length measured. Thermal expansion can also be measured with a high temperature X-ray diffractometer where the change in lattice spacing along different lattice planes is measured. Ceramics show a wide range of thermal expansions, over better than a decade. Alkaline earth oxides have high expansion coefficients. The coefficient is the expansion % over one degree of temperature rise. Fused silica glass has a very low coefficient that explains why it is used in dilatometers as the sample holder and push rod. Single crystal sapphire is often used for higher temperatures.

Thermal expansion of zirconia was previously discussed and varies depending on the amount of stabilizer. The transformation of the monoclinic phase to tetragonal is a contraction and reduces expansion as its concentration increases in the cubic matrix.

Figure 11.11 is of a dilatometer that can be routinely used for thermal expansion measurements.

Loss on Ignition (LOI)

LOI is commonly reported by chemists as part of an analysis. It is simply the amount of weight loss after the sample is heated to a red heat. The temperature is sufficient to dehydrate and burn off organics.

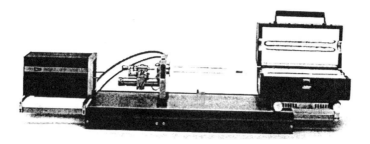

Figure 11.11: Dilatometer. Measures thermal expansion and glass transition temperature for glasses. (Courtesy Theta Instruments)

Check List, Thermal Analysis

- TGA
  Weight loss on heating
  - Atmosphere
  - Binder burnout
  - Desorption
  - Decomposition
- DTA
  - Phase changes
  - Decomposition
  - High temperature reactions
- Thermal Conductivity
  - Farm measurements out.
  - Materials can have different conductivities in different directions
  - Some materials are selected for their conductivity
  - Porosity reduces conductivity
  - Fibers are useful as thermal insulation
- Specific Heat
  - Not usually measured
  - Need for thermal conductivity measurements

- Thermal Expansion
    - Usually measured with a dilatometer
    - Low expansion improves resistance to thermal shock
- LOI
    - Routine with chemical analysis
    - Often part of materials specifications

Particle Size

Measurement of particle size distribution is key for characterizing a ceramic powder. This data is often presented as a plot of *% Finer Than, with the logarithm of Particle Diameter*. A typical plot is shown in Figure 11.12.

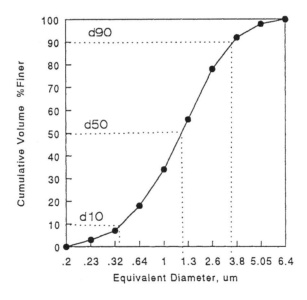

Figure 11.12: Particle Size Distribution Plot. Data is usually presented in this form.

When the curve is steep, the distribution is tight. When the curve is shallow, the distribution is broad. These results are often summarized by three values: d_{10}, d_{50}, and d_{90}. The two end values are sometimes instead selected as d_{20} and d_{80}. These are marked on the preceding figure, where d_{10} indicates that 10% of the material is below the size indicated, which in this case is about 0.38 μm.

Light Diffraction/Dynamic Light Scattering (DLS)

Coarse particles are best measured by screening, as previously described. Fine particles are usually measured by light diffraction or DLS. Other techniques are based on settling either with or without a centrifuge. Doppler shift measures very fine (to 0.003 μm) particles due to Brownian motion. A number of good general purpose instruments exist, one of which is shown in Figure 11.13.

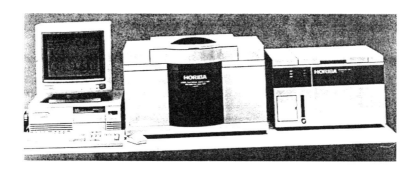

Figure 11.13: Particle Size Analyzer. Several types of instruments are available. This one uses light diffraction. (Courtesy of Horiba)

Analysis is performed by taking an aliquot of the material, adding surfactant, and dispersing ultrasonically with stirring. A magnetic stirrer is used, with a top surface having an angle that lifts the suspension vertically. The sample, in a beaker on a lab jack, is encased in a soundproof chamber. The sample is sonified at a set amplitude for a specified period of time, determined experimentally. After sonification, a few drops of the suspension are removed using a pipette. Using throw-away polyethylene pipettes is convenient. Drops are added to the instruments reservoir until the desired concentration is attained, as described in the operating instructions. Measurements are then made and printed out.

While flat-ended horns are best, they abrade and must be remachined. Only a minimum of material should be remachined as machining renders the horn out-of-tune. Since the metal is titanium and difficult to machine, cutting is best done using a sharp-cemented carbide cutter at a moderate cutting speed with coolant.

When engaged in process research, there exists the need for real-time data. Measurements might be needed in a few minutes rather than the few days that is typical of an analytical lab. Such time constraints necessitate an on-site instrument.

Since dust fogs up the optics, install a blower and air filter on the instrument to keep the optics clean. Ensure that the air flow is from the lab through the filter into the instrument, and not the other way around.

Microscopic Particle Size Measurements

An alternative to light diffraction is microscopy. Colloidal-sized particles are better visualized with a TEM than by light diffraction. Additionally, particle shape can be determined. Light diffraction simply provides an "Equivalent Particle Diameter." The principle disadvantages are that microscopy is much slower and more laborious. However, microscopy is direct and precise. Microscopy should be intermittently used to check for foreign contaminants and particle shape. SEM is a good tool here as is optical microscopy when the particle size is larger than a few micrometers.

Preparing a sample of powder for SEM is tricky because particles agglomerate. Settling and drying onto the stub of a low molecular weight organic, such as methanol, helps to retard agglomeration. The suspension should be diluted in order for individual particles to be visible. Good and poor dispersions are illustrated in an SEM photo of alumina particles. See Figure 11.14.

With poor dispersion, individual particles are not separated to the point of observation or measurement.

When using an optical microscope for larger (3 μm+) particles, prepare a slide by placing a small amount of a ceramic powder onto a glass slide, adding a drop of an index of refraction liquid, and placing a cover glass over the sample. Press down the cover glass and smear it around to disperse the powder using a pencil eraser. When there is a wide variety of particle sizes in the sample, the material can segregate with coarse particles moving toward the edges. Watch for this. Excess liquid produces a mess. Excess powder renders the slide opaque, making visualization of particles difficult. With water's low viscosity, micrometer-sized particles will jump around by Brownian motion. Using a more viscous oil eliminates this problem. Sets of index of refraction liquids are available for immersion of powdered samples. View the powder with plane or polarized light at the appropriate magnification, examining for size, shape, color, transparency, contrast with the index liquid, and birefringence.

Most ceramic systems have only a few phases present. They are more easily characterized by SEM compositional analysis, particle size analysis instruments, and X-ray diffraction structural analysis. Ceramic microscopy of powders is thereby simplified. A quick glance at a slide is by far the easiest way to obtain an impression of the size, shape, color, transparency, contrast, and birefringence of the ceramic powder.

Contrast for colorless ceramics is determined by the difference between the index of refraction of the ceramic and the index liquid in which it is immersed. Unlike the story of the jewel thief who hides diamonds by dropping them into a glass of water, the ceramic particles become invisible in plane light when the two indices are identical. The jewel thief is doing time since the index of water is 1.333 while the index of diamond is 2.417 a difference of 1.084. The mineral cryolite has an index of 1.338, close to that of water, and can be concealed in a glass of water, though such information might prove worthless to a precious-jewel-seeking thief.

Figure 11.15 shows alumina in two different index of refraction liquids.

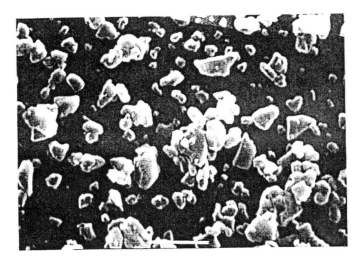

Figure 11.14: Quality of Powder Dispersions. The two figures show a good and poor dispersion of alumina powder. SEM Photographs. Scale bar 1.0 μm.

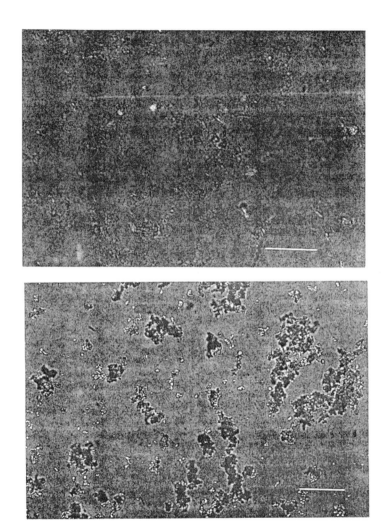

Figure 11.15: Contrast By Index of Refraction. A modest difference between the index of the powder and the liquid is needed to see the particles. Transmitted light slide. Scale bar 100 μm.

The picture on the bottom is in a liquid of 1.610 refractive index. Alumina has an index of about 1.76. Due to this large difference, the particles are readily visible. The picture on the top is in an index of 1.76. Although the concentration of particles is about the same, they are almost invisible. Therefore, a ruby in a glass of this index liquid has virtually no contrast. However, the ruby is red and the jewel thief finds himself once again in the slammer. Color is another way to discern the differences between phases.

It is very easy to take a quick look at a ceramic powder on a slide and discern between phases by index, shape, size, or color. With two phases present, when the mixture in an index liquid matches one phase, the other is visible by itself due to contrast. The technique has ppm sensitivity and is easy to do.

Astrophysicists are engaged in a conspiracy against petrographers with their claim that the speed of light is a universal constant, in all directions and regardless of the motion of the light source. What they choose to hide is that the speed of light depends upon the index of refraction of the medium through which it is passing. High index equals low speed. And, the speed of light is not the same in all directions through anisotropic crystals. Those include the multitude of all crystals except cubic. Differences in light speed in the crystal throw the light out-of-phase in polarized light with interference. This is called birefringence and is visible in a powder or thin section when the polaroids are crossed. Figure 11.16 is of two powders where one has a much higher birefringence than the other. This causes it to appear much brighter in the field of view.

The bright particles are zircon where the index in one crystal direction is around 1.923 and in the other 1.968, with a difference of 0.045. The other phase is alpha alumina (corundum) where the indices are 1.767 and 1.759, with a difference of 0.008. Zircon is said to have a high birefringence while alumina is said to have a low birefringence. Zircon is easily spotted in the slide because it is much brighter in polarized light than alumina. The test between two materials of different birefringence is very sensitive, and can easily be in the ppm range just with a glance.

When the index of refraction of the liquid is close to that of the powder, the particles disappear unless they are colored or contain porosity. That is a good way to determine the index of the ceramic, which is one of its diagnostic criteria. Alumina and mullite have an index of refraction of about

1.74 to 1.76. For good visualization, the index of the liquid should be about 1.6 to provide some contrast. When the indices are far apart, the contrast will be so high as to render the particles almost opaque. It is advisable to use a liquid with an index closer to that of the ceramic. Always use a cover glass to prevent damage to the objective lens.

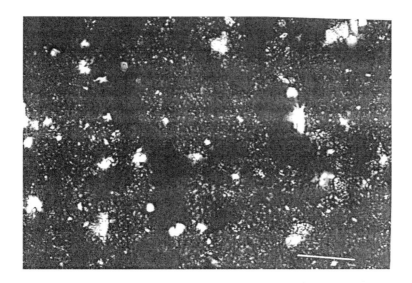

Figure 11.16: Contrast by Birefringence. When two or more materials are present, the birefringence can distinguish between them. Transmitted light slide. Scale bar 100 μm.

Check List, Particle Size

- Characteristics of the distribution curve
- Light diffraction
    Ultrasonic dispersion
    Stirring
    Prevalent technique
    Fast

- SEM
    Particle shape
    Foreign materials
    Stub preparation
    Magnification, 20,000X ± max.
    Particle count labor intensive
- TEM
    Great detail
    Labor intensive
    200,000X ± max.
    Particle count labor intensive
- Optical Microscope
    Slide preparation
    Particle shape
    Optical properties
    Particle count can be labor intensive 1000X, max.

Density Measurements

The density or specific gravity of a sample is an essential measurement. The two cases of permeable and impervious materials are discussed below.

Permeable Materials

Measurement can be performed in two ways. The surface can be sealed and the specific gravity determined. Or, the sample can be saturated with water with the bulk density and apparent porosity measured.

Sealing the surface. Perform the following steps:
- dry the sample and cool,
- weigh,
- attach a thin suspending wire or cord that can be weighed,
- quickly dip in liquid paraffin just hot enough to be fluid, and
- weigh suspended in water. Brush off bubbles.

Using the specific gravity of the paraffin and the layered amount, one can correct the buoyancy and determine the apparent specific gravity of the sample. The wax should not penetrate the pores. This is why one uses a temperature where the paraffin is close to the melting point so that it will freeze quickly. This is also why the dip is quick so that the sample does not have time to heat up.

Saturating the sample with water. The ASTM method involves boiling in water to fill the open pores with water. There is a better way. Figure 11.17 is a chamber for evacuating the air from the open pores in the sample, covering with water, and then applying pressure to force the water into the pores.

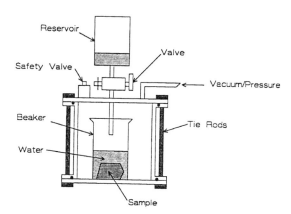

Figure 11.17: Vacuum/Pressure for Saturation. The apparatus is used to infiltrate the open porosity with a liquid.

The procedure is as follows.
- Clean, dry, cool, and weigh the sample.
- Place the sample in a beaker that is then put into the vacuum deairing apparatus, close it, and a draw a vacuum.
- Water with a drop of a wetting agent then enters the beaker through the valve on the top of the lid. Use enough water to cover the sample. The water will splatter in the vacuum due to air bubbles, so a tube, long enough to be below the beaker's rim, is attached to the bottom of the lid.
- Either pressure is applied to the chamber or, if not required, air is admitted and the atmospheric pressure forces the water into the pores. Use additional pressure when the sample is difficult to infiltrate with atmospheric pressure alone.
- The sample is now saturated. Blot the sample surface and weigh it.

Measurements are made using a balance shown in Figure 11.18.

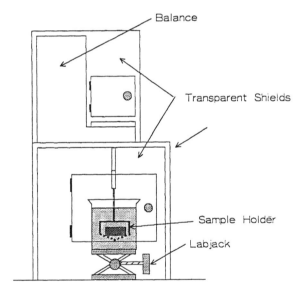

Figure 11.18: Balance for Specific Gravity Measurement. The balance must have a means to connect the pan to the lower chamber.

Density Measurement. For small samples, the balance is a standard analytical balance with a way of connecting a sample holder to the weighing mechanism through the bottom of the balance. The sample holder screws into a connection to the balance pan. The balance is supported on a stand that has a transparent breeze shield and a door. The lab jack is used to raise and lower the sample holder, which, in this case, is a spiral of stainless wire basket with a connector for attachment. The procedure is as follows.

- Attach the empty sample holder to the balance and submerge it in a beaker of water containing a drop of a wetting agent up to the mark on the connecting rod and tare the balance.
- Disconnect the sample holder from the balance pan.
- Then quickly place the sample on the submerging sample holder and place it in a beaker of water containing a drop of a wetting agent.
- Tip or swish it to remove bubbles.
- Raise the beaker with the labjack and connect the sample holder to the balance keeping the sample submerged.
- Raise the beaker again to where the surface of the water is at the mark on the connecting rod.
- Now weigh the submerged sample.

The following equations are used for the calculations.

$$\text{Bulk Density} = a / (b - c) \tag{11.1}$$

$$\text{Apparent Porosity} = (b - a) / (b - c) \tag{11.2}$$

$$\text{Apparent Specific Gravity} = a / (a - c) \tag{11.3}$$

Where:
a = dry weight
b = saturated weight
c = suspended weight

Impervious Materials

Since there is no need for saturating the sample, those steps can all be eliminated. When unsure, the saturation procedure provides the answer when calculations are made.

The procedure is essentially the same, expect the infiltration steps are avoided. As these samples are often small, the procedure has some additional steps to impart more precision.

- Handle the sample with tweezers after it has been cleaned.
- Wait until the weight swing settles down before recording it.
- Measure the temperature of the water and make a correction for the water density. Handbooks on chemistry have these tables.
- Specific gravity is simply a / (a-c).

Mercury densitometers are available. Mercury is poisonous and must be handled with great care. The poison is cumulative over many years and has devastating effects on brain function. For these reasons, I do not recommend a mercury densitometer but instead advocate finding alternatives. For example, one can machine the part into a geometrical shape and weigh it.

Check List, Specific Gravity

- Permeable Materials
  - Paraffin procedure
  - Saturation apparatus
  - Saturation procedure
  - Use of vacuum/pressure
  - Measurement procedure
  - Calculations.
- Impervious Materials
  - Saturation is not necessary.
  - Measure the water temperature and correct for its specific gravity.
- Measuring Procedure

3.0 SLIP PROPERTIES

Some slip properties were discussed in Chapter IV on slip preparation. This section provides additional detail on viscosity measurements. Three types of viscometers are described.

Strain Rate Controlled Viscometers

This type of viscometer usually has a rotating element submerged in the slip where the speed of rotation is controlled and the viscosity or stress is measured. Viscometers measure viscosity. Rheometers measure strain rate and shear stress where a stress/strain curve can be obtained. The principle difference is in the software and peripherals. Figure 11.19 is a picture of a common viscometer.

The instrument is digital, with controls for selecting shear rate and rotor speed (which is related). The spindle type is selected to put the measurements on scale. Slip temperature is also measured.

As with most modern instruments, software, a computer, printer, and accessories are available. Accessories include a spiral adapter that can be used for pastes and thixotropic materials, a helipath stand that is even better for thixotropes, temperature controls, and means to prevent evaporation of the sample. Some data taken with an instrument is shown in Figure 11.20.

Measurements are first made by increasing the spindle rpm in increments and then decreasing it with the same settings. The data in the figure shows pronounced shear thinning where the *up curve* follows the relationship:

$$\text{Log viscosity} = -0.886 \log \text{rpm} + 4.1495 \qquad (11.4)$$

$$\text{Correlation coefficient} = -0.9998 \qquad (11.5)$$

The *down curve* follows the relationship:

$$\text{Log viscosity} = 0.895 \log \text{rpm} + 4.17 \qquad (11.6)$$

$$\text{Correlation coefficient} = +0.99995 \qquad \textbf{(11.7)}$$

This very high correlation coefficient indicates an almost perfect correlation. Data this precise instills confidence.

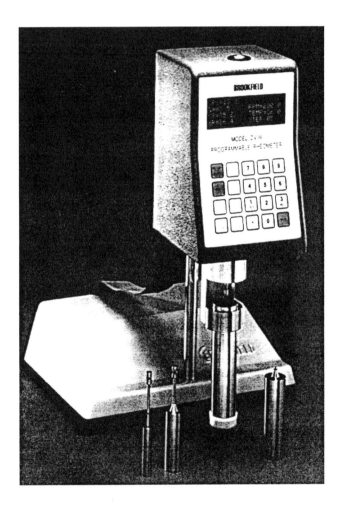

Figure 11.19: Viscometer, Strain Controlled. Rotor speed is varied and the stress measured. (Courtesy of Brookfield)

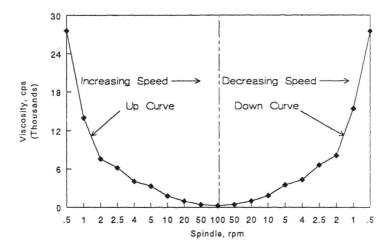

Figure 11.20: Viscosity/Shear Rate. The up curve is on the left and the down curve on the right of the diagram.

Spindle

Select a spindle with enough drag to put the measurements on midscale. As expected, large rotors have more drag than small ones. Typical configurations are shown in Figure 11.21. (Courtesy Brookfield)

Sample viscosity is instrumental in rotor selection. With experience, one can select the right rotor after one or two trials. Keep the rotor clean and handle it with gloves. Discard the rotor when it becomes dinged or bent since it will produce false readings. Rotors thread onto the shaft with a left-handed thread; keep this in mind to avoid torquing it the wrong direction and damaging the innards of the instrument.

Figure 11.21: Viscometer Rotors. The rotor is selected to place the measurements on scale. (Courtesy of Brookfield)

Rotors should be inserted into the slip at an angle to prevent entrainment of air bubbles. Then, thread them onto the shaft. Viscosity usually changes for a time as the rotor starts to revolve so it is a good idea to wait for the measurement to stabilize. This does not apply when using the spiral adapter for thixotropic measurements. In this case, avoid disturbing the slip at the slowest speed and watch the readings as they quickly maximize. Then turn to rotation off at once.

Figure 11.22 is of a spiral adapter attachment that can be used for pastes and for approximate thixotropic measurements.

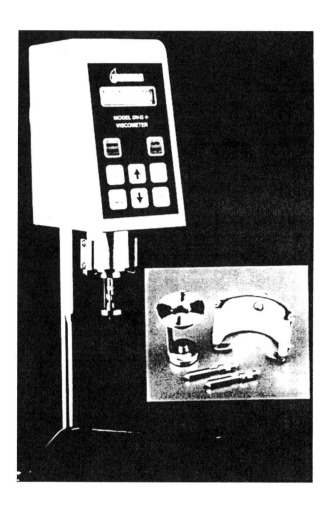

Figure 11.22: Spiral Adaptor. Useful for thick pastes and can be used for thixotropic measurements. (Courtesy of Brookfield)

Figure 11.23 is of a thixotropic slip measured with the spiral adapter. A wet cloth covers the slip to prevent evaporation.

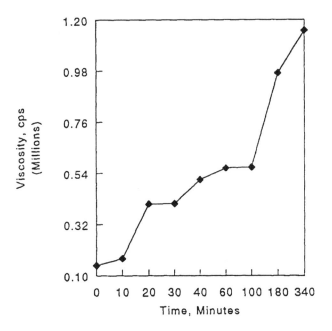

Figure 11.23: Thixotropic Data. There is a very large increase in viscosity over time.

The data jumps around somewhat but the thixotropic nature of the slip is clearly seen, with the viscosity rising from about 0.1 to 1.2 million cps over a period of 340 minutes. The helipath stand is even better for thixotropic data. This stand has a drive that slowly raises the rotor up through the mix, thereby supplying fresh slip with a minimum of disturbance. Run the test by taking a measurement and turning the drive off for an interval depending on the rate at which the slip develops thixotropy.

Stress Controlled Viscometers

In a stress-controlled viscometer, the rotor speed is set by the instrument based on the stress it senses. Usually, this instrument is used for fine-grained slips with a cone and plate sample holder. Plate and cone configurations require a very small sample of slip. This can be advantageous in some instances, such as when the slip or paste is comprised of precious metals. Figure 11.24 is of a stress-controlled viscometer.

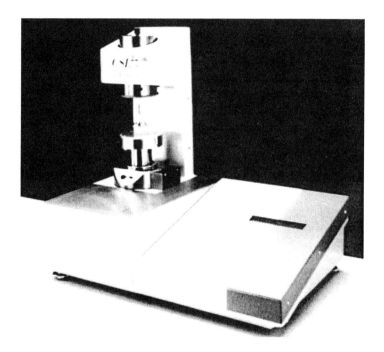

Figure 11.24: Viscometer, Stress-controlled. Stress is set and the strain necessary to attain that stress is measured. (Courtesy of TA Instruments)

The instrument has accessories for controlling the temperature and an enclosure for inhibiting evaporation of the slip's liquid. Enclosures are especially important when the test interval is long or the liquid has a high vapor pressure. Of particular importance is the capability for measuring viscosity at very low strain rates. Data at low strain rates provides information on processes such as leveling, sag, and behavior at slow flow rates. This type of information cannot be extrapolated from higher strain rate data as the viscosity often departs from the extrapolation path.

Orifice Flow Viscometers

These are simple cylinders with a hole in the base. The test is run by filling the tube with slip, opening the hole, and measuring the time it takes for the stream to break. This number is valid when calibrated with standards, but no information is available at various strain rates. The Zahn viscometer is of this type and is useful for quality control when other equipment is not available. The device is inexpensive and thus is advantageous for the tightly-budgeted. Figure 11.25 is a sketch of this apparatus.

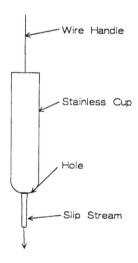

Figure 11.25: Zahn Cup Viscometer. Useful for QC in the plant.

Zeta Potential

Zeta potential is measured by the velocity of the ceramic particles in an electric field. This is often accomplished by direct observation, but becomes difficult if more than one phase is present. Given that a ceramist ordinarily works with only a few materials at a given time, it is easier to send the samples out for measurement. One inexpensive instrument has been around for a while but, unfortunately, does not work. Be sure that instruments are always demonstrated on your materials before purchasing.

pH

To use a pH meter, simply follow the instructions on the box. Keep the instrument calibrated, and keep the electrode clean and wet in the storage solution. Dried-out, dirty electrodes are the principle cause of instrument failure.

Check List, Slip Properties

- Strain Rate Controlled Viscometers
  Spindle type
  Spindle speed
  Measure up and down
  Measurement procedure
  Curve fit
  Thixotropic: spiral adaptor, Helipath
- Stress Controlled Viscometers
  Plate and cone
  Accessories: temperature control, evaporation control
- Orifice Flow Viscometers
  Lacks strain rate data
  QC
  Inexpensive

- Zeta Potential
    Usually better to farm work out
- pH: Care for Electrodes

4.0 MICROSCOPY

This section discusses both optical and electron microscopes. Both microscopes have their particular limitations and advantages. While there are a wide variety of optical microscopes, those usually encountered in the ceramic lab are either stereo-binocular or petrographic/metallographic. Most likely, the microscope is a universal microscope, having both transmitted and reflected light capability. The various uses of microscopy include particle size measurement, dimensions, grain size, microstructure, morphology, phase identification, stress measurement, chemical composition (electron microscopes), and crystal type identification (TEM). TEMs have also been used successfully for defect structure analysis.

They are also used to examine surfaces of fired ceramics for cracks, chips, spots, interfaces and other properties.

Many of the attributes for study are too small to be observed without magnification. But microscopes do more than just magnify, as seen in the foregoing partial list of uses.

Optical Microscopes

This section describes two types of optical microscopes, stereo-binocular and universal. Many other types exist for specialized uses, and there are many other instruments that use optical microscopes as part of their design. Because the stereo-binocular and universal microscopes are most common in the ceramics lab, the discussion will be limited to these.

Stereo-Binocular

These are commonly used to look at the surface of ceramics and other materials where the view is similar to that produced by a magnifying glass, but at magnifications up to about 40X. Illumination is oblique; the light source illuminates the surface without going through the optics. These common microscopes are quite useful. Flexible fiber illuminators with an intensity control are preferred. Photographic capability is also useful for specialized cases. Being stereo, there are two objective and two ocular lenses. This is a substantial advantage over single lens configurations as depth perception helps in visualizing the surface and, perhaps even more importantly, prevents eye strain. Ocular focus is adjustable by turning a sleeve on the lens. High eye point oculars allow the ceramist to wear glasses while using the microscope. Focus allows individual adjustment of the oculars to suite one's eyes.

To focus both eyes, first focus on the stage with one ocular, using the rack and pinion. Using the other ocular, turn the sleeve on the lens to where the same spot is in focus. Each individual must adjust the lens for his/her eyes.

It is unwise to purchase a microscope with sub-standard optics since its use will be limited.

A stand with an extended arm is useful for observing large objects. Ordinary stands are limited to about 100mm working distance, limiting the sample size that can be observed.

Zoom is available as standard equipment but is not essential. However, three choices of magnification are useful. When zoom is included, ensure that the sample stays in focus as the magnification is changed. If the focus changes, try another brand.

Figure 11.26 is of a stereo-binocular microscope.

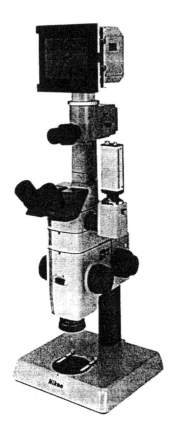

Figure 11.26: Stereobinocular Microscope. Very useful for looking at sample surfaces up to about 40X. (Courtesy of Fryer-Nikon)

Universal Microscopes

Universal microscopes are usually binocular and equipped for transmitted and vertical illumination, both of which can be polarized. Like the stereo binocular, oblique illumination can be used to moderate magnifications of about 100X. This section discusses all three light sources. Useful magnifications are up to 1000X with oil immersion objectives. Smaller objects necessitate electron microscopy. Figure 11.27 is of a universal microscope.

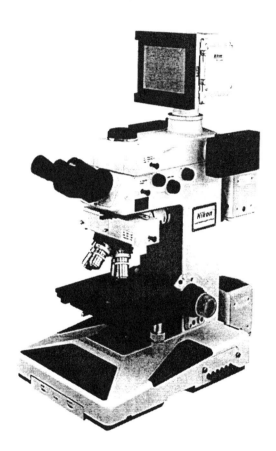

Figure 11.27: Universal Microscope. Used for reflected, transmitted, and oblique illumination. (Courtesy of Nikon)

Current, well-designed microscopes have a built-in light source, automatic photographic capability, and, most importantly, a flat, clear field of view. Discussed below are different illumination techniques.

Vertical Illumination. In this configuration, the microscope is similar to a metallograph where polished sections of the samples are prepared. The light source passes through a vertical illuminator, through the objective lens, reflects off the polished surface, reenters the objective, and travels into the ocular lenses. Options include either one or two oculars; eye strain is decreased with two oculars.

The vertical illuminator contains the polarizer, another polarizer called an analyzer, and two iris diaphragms. One controls the light intensity, and the other controls the field of view. This diaphragm is set to where the field of view is just a little larger than that which is observed. This cuts down light scatter which makes for a foggy field. To proceed, close the diaphragm, center the spot with the levers provided, and then just open it up to the edges of the field. When viewing is really foggy, the field can be closed down to where just part of it is viewed. Another way to reduce light scatter is to sputter-on/vacuum evaporate a thin reflective metal film onto the polished surface. The film is far too thin to obscure any detail. Figure 11.28 is of a polished section with the surface coated with a gold film in one case and plain in the other.

The difference in obtaining a clear field of view is evident from the two photos. This is also a useful procedure for measuring microhardness indentations. The ends of the indents are much sharper. Now visible, the indent is always longer and the microhardness softer, but really more accurate.

Many instruments have Nomarsky interference contrast that imparts color to the viewed surface depending upon the relief. Relief is the difference in altitude of the surface. Hard phases stand out of the surface while softer phases are in slight depressions. This relief is easier to see if they show different colors. Nomarsky also sharpens etched grain boundaries and other features, as it imparts color to the field of view.

As an example, the slip-cast ceramic contains an alumina grog that is relatively coarse, and a finer bonding phase made up of alumina and glass. Figure 11.29 is a polished section of this material taken near the top of the casting observed with Nomarsky illumination.

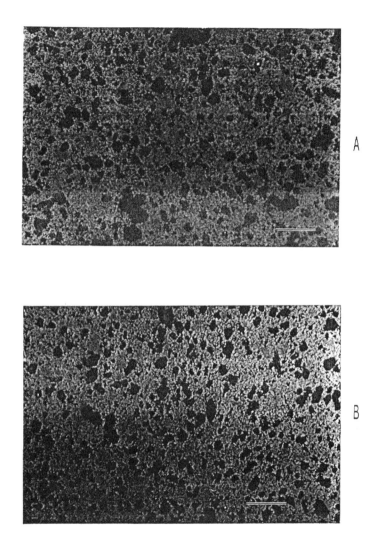

Figure 11.28: Light scattering on an alumina polished Section. The two views show how metallizing the surface reduces scattering. Scale bar 100 μm.

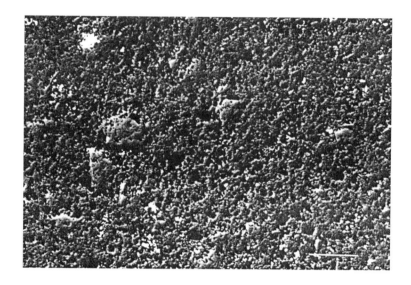

Figure 11.29: Slip Cast Coarse Grained Alumina Ceramic, near top. Polished section. Polarized light, Nomarski interference contrast. Concentration of fines at the top of the casting. Scale bar 200 μm.

Color can be adjusted depending on how much of the Nomarsky is added. In this case, blue brought out the distinction between the coarse grog and the finer-grained matrix, a matter of choice. Measure grain size by the grain intercept method. The concentration of large particles, small particles, and porosity can also be measured on this surface and is quantitative.

Another section taken near the base of the casting is seen in Figure 11.30.

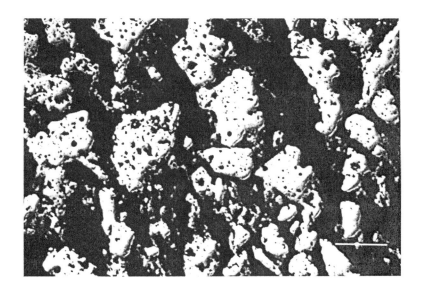

Figure 11.30: Slip cast Coarse-Grained Alumina Ceramic, near base. Polished section. Polarized light. Nomarski interference contrast. Large particles settled out during slip casting. Scale bar 200 μm.

The microstructure is entirely different. Coarse particles have settled out as the piece was cast, leaving the top too fine and the bottom too coarse. Firing shrinkage will be different as the fine top will shrink more than the coarse base. Such circumstances induce distortion and cracking.

Before leaving this example, it is useful to discuss what can be done to bring this serious segregation of particles under better control. Take a look at Stokes' law,

$$V_t = (2g (d_1-d_2)r^2) /9n \qquad (11.8)$$

where V_t is the terminal velocity of a spherical particle, g is the acceleration due to gravity, d_1 is the density of the sphere, d_2 is the density of the liquid, r is the spherical radius, and n is the viscosity of the liquid. In order to slow down segregation, V_t should be minimized.

Reducing "g" casting would require an orbit. That is not practical.

"d_1 an d_2" are fixed by the systems composition.

"r" can be reduced if other properties of the ceramic are not seriously compromised. This has a large effect as the radius is to the power of 2.

"n" can be increased with thickeners such as long chain polymers or mineral thixotropic agents such as Hectorite. Polymers sometimes can interfere with the casting process or the plaster, but they are used. Hectorite is effective in very small (1% or less) amounts to increase viscosity.

"V_t" Any velocity is a function of both distance and time. Time can be cut substantially by pressure casting. That solution may be best as it does not interfere with the composition or surface chemistry.

See what can be learned from a polished section!

Measuring the grain intercept was described earlier. Random lines across the field are used to locate the grains or pores that are measured, obtaining the mean intercept and the standard deviation. When more than one phase is present, the volume percent of each phase is determined quantitatively. When only a rough idea of the grain intercept is necessary, one can visually scan the field and pick and measure one grain that appears to be average.

If the sample were comprised of spheres, there is a mathematical way to convert grain intercept to grain size. Since the ceramic sample does not have spherical grains, estimates to make the correction do not add to the information but rather subtract from it.

Oblique Illumination. This viewing method is where the illuminator shines directly upon the top of the specimen and does not go through the vertical illuminator. The field of view is "natural" such as that of a magnifying glass or stereo-binocular microscope. A flexible fiber optic illuminator is especially useful as it can be moved near the end of the objective close to the sample surface. The technique is good to about 100X, above which it is difficult to illuminate the surface as the clearance is too tight. Figure 11.31 is a view by oblique illumination of the surface of vacuum-fired porcelain enamel.

Figure 11.31:   Vacuum-Fired Porcelain Enamel Defect. Oblique illumination is useful for viewing surfaces. Scale bar 400 μm.

An interest in vacuum firing derives from pin holes that sometimes penetrate porcelain enamel. By vacuum firing, the porosity can be eliminated. Ordinarily, porcelain enamel has a bubble structure that is essential to its structure. The figure shows that the bubbles have indeed been eliminated by the vacuum firing. This makes the coating extremely brittle. In the figure, the blue background is due to cobalt in the ground coat. Specks are mineral remnants.

The large central figure is where a large bubble emerged through the enamel coating. These bubbles derive from the decomposition of water on the steel surface.

Hydrogen ions can then penetrate the steel, pass through an iron-rich layer at the interface, and then bubble up through the porcelain layer. Liquid glass is carried along with the bubble and an iron-rich phase precipitates out during cooling, creating the central feature in the figure.

Porcelain false teeth are vacuum-fired to remove porosity; this doesn't work well for porcelain enamel coatings.

Gases sometimes penetrate the steel substrate and form bubbles that then rise through the glass, carrying with it the iron-rich layer next to the steel. In the trade, these are known as *copper heads*. This is a view using oblique illumination. The structure and its implications would not have been evident by other viewing techniques.

Transmitted Illumination. In transmitted illumination, light is reflected up through the substage, through the slide containing the sample, into the objective lens, through the upper polarizer, and then into the ocular lenses. The substage contains a diaphragm controlling the light intensity, a second diaphragm controlling the field of view, a polarizer, and centering screws.

To center the field, close down the diaphragm and center using the screws. The objective lens also has centering screws that are used as the stage is rotated to make the substage coaxial with the objective. The oculars are focused in the same manner as described in the discussion on reflected light.

Since light is passed through the ceramic, there is much more information available than when it is simply reflected off its surface. Optical petrography is a fairly complex subject, much of which is not usually germane to the study of ceramics. With ceramics, the phases present are usually known and have a much more limited chemical constitution than the extensive array of minerals found in nature. After all, ceramics usually are constituted from only a few light elements. Other methods of analysis have largely supplanted optical microscopy for the identification of the phases present. The combination of X-ray diffraction and chemical analysis by a spectrographic method usually pinpoints the identity of the phases present. But, X-ray diffraction and an analysis do not reveal much about microstructure, size, or shape. This is where optical microscopy is useful.

Transmitted light microscopy is applicable for examining films, powders, and thin sections. Films are usually polymers and not very applicable for ceramics. Powders are routinely examined by optical microscopy whenever the particle size is above about 2-3 μm. A slide is made by placing a small amount of powder on the slide and then adding a liquid, which is usually an index of refraction liquid. Mix the particles to disperse them and then put on a cover glass. Always use a coverglass to create a flat surface and to protect the objective lens. A rubber eraser is then used to move the cover glass around to distribute the powder suspension evenly and to squash the soft agglomerates.

When the particle size distribution is large, the sample segregates on the slide with the coarse particles moving toward the edge. Watch for this and make a new slide if it occurs. If water is used as the liquid, the small particles exhibit Brownian motion due to the low liquid viscosity. In order to see the particles, contrast is needed between the field and the transparent particles. This contrast is controlled by the difference in the index of refraction between the samples particles and that of the liquid. A big difference yields the most contrast.

Slides of particles are useful for measuring particle size and shape. Eyepieces are often fitted with a reticule or, even better, a movable scale called a *filar*. The filar is calibrated with a stage micrometer that is a precisely ruled slide. This yields an eyepiece calibration where a movable cross hair can be moved from one side of a particle to the other, the difference being the particle size. A revolving stage is recommended where the slide can be rotated, say 90 degrees, and the other dimension measured, resulting in some information on particle shape.

The microscope can also be fitted with an accessory called a mechanical stage. A mechanical stage is attached to the revolving stage and has rack and pinion movements where X/Y locations can be recorded. It is useful to be able to return to a previous location on the slide, especially when working at high magnifications, in order to place the cross hairs at exact locations. It is difficult to move the slide by hand with this precision. A mechanical stage is recommended.

Thin sections are slices cut from a sample and ground thin enough to become transparent without overlapping grains. Some considerations for making sections are addressed later. Obviously, thin sections are used with transparent ceramics. The birefringent color in polarized light depends on the thickness, with thicker sections showing higher birefringent colors than those ground more thinly. Residual stresses in ceramic surfaces due to grinding make it difficult to thin to petrographic standard thickness. The specimen spalls during preparation due to these stresses. Having a slightly thick section is preferrable to having no section at all. For dense ceramics, a thickness of 30-40 μm is sometimes achievable. Grain size must be larger than that or there will be overlapping. One compromise is to stop lapping when the edge is thin enough, leaving the center too thick.

Zirconia ceramics are especially amenable for study with thin sections because the monoclinic phase is birefringent while the cubic phase is not. Figure 11.32 is of a thin section viewed in polarized light.

Figure 11.32: Microstructure Magnesia Stabilized Fine Grained Zirconia, Thin Section. Polarized light.   Grain boundaries stand out due to birefringence. Scale bar 200 μm.

Cubic grains are isotropic and appear dark. Grain boundaries in this section are a little birefringent, containing some of the monoclinic phase and probably some impurities.

Alignment of the grains in the microstructure is not visible except in thin sections observed in polarized light, providing that the phase is birefringent. As the stage is rotated, a grain becomes extinct. Extinction occurs when the birefringence becomes zero and the grain darkens. When all or most of the grains become extinct together, there is alignment in the microstructure. For more sensitive viewing, a gypsum plate is inserted into the optical path between the polarizers. This imparts a background red color that accentuates the differences in orientation. The gypsum plate is called *first order red* or the *sensitive tint plate*. Figure 11.33 is a thin section of fused cast alumina near the interface with a steel casting plate.

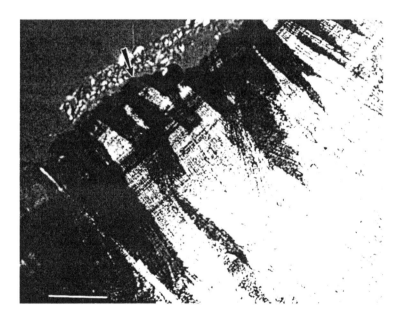

Figure 11.33: Fused Cast Alumina. Thin section, polarized light, first order red interference plate. The edge on the right was the interface against the steel mold. Scale bar 400 μm.

Alignment is a consequence of crystallization which is dendritic under these conditions. Contiguous regions of the same color have the same crystallographic orientation. Bundles of blue dendrites are attached to the same crystal lattice, as are bundles of red or yellow dendrites attached to each of their separate crystal lattices. These bundles are interfaces that deflect crack propagation and impart toughness to the ceramic. As an additional point of interest, the dark bands paralleling the cast surface on the right are waves of impurities pushed out ahead of the advancing crystallization front. Processing can also orient grains in a ceramic structure.

Shear orients platy or elongate particles because the shape is often related to the crystal structure. Thin sections are one way of observing alignment.

Due to the photoelastic effect, stresses can also be seen in thin sections. In particular, stresses around pores in zirconia result in dark bands around the pore as a void acts as a stress riser. This is rare with other ceramics due to lower photoelastic constants or less stress in the structure.

Zirconia refractories, where thermal shock is severe, are formulated from coarse monoclinic zirconia grains bonded together with a partially MgO-stabilized matrix. Figure 11.34 is a thin section of this material viewed with transmitted polarized light, and with the sensitive tint plate inserted.

Figure 11.34: Magnesia Stabilized Zirconia Refractory, Thin section, polarized light, first order red interference plate. The structure shows twinned large monoclinic zirconia grains. Scale bar 400 μm.

The field is largely covered by two large monoclinic zirconia grains. The matrix can be seen around the two large grains. Due to the large volume change when tetragonal zirconia transforms to monoclinic, the grains are extensively twinned.

On the same thin section, another location shows a concentration of the matrix, seen in Figure 11.35.

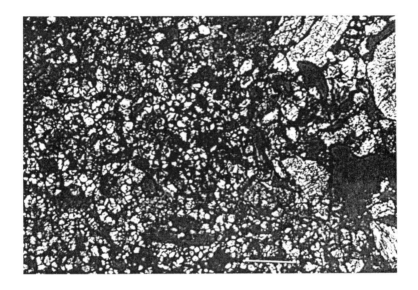

Figure 11.35: Magnesia Stabilized Zirconia Refractory (same sample as previous figure). Thin Section, polarized light, first order red interference plate. Fine particles are not mixed well into the body. Scale bar 200 μm.

As there is a concentration of the bonding phase in this location, there must be a sparsity of bonds in others. There was a mixing problem that resulted in a refractory structure off specification. Mixing often leads to problems, as the fines with a much higher surface area tend to stick to themselves. There are two ways to obtain a better homogeneity. One can either increase shear to break the agglomerated fines or one can pre-wet the coarse grain with the binder solution while mixing sift in the fines. This is a bit tricky and requires practice. Either a polished or thin section provides the measure of how well mixing has been accomplished.

There is another location on the same thin section where another problem is seen, as shown in Figure 11.36.

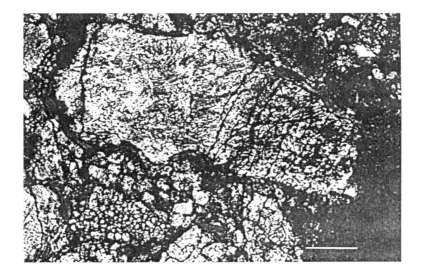

Figure 11.36: Magnesia Stabilized Zirconia Refractory (same sample as previous two figures). Thin Section, polarized light, first order red interference plate. The bond has shrunk away from the large grain. Scale bar 100 μm.

Around the large grog particle is an almost-continuous void where, during sintering, the matrix pulled away from the grain as the matrix shrunk. During sintering, when there is little liquid present, the particles must be in physical contact in order to bond. Atoms cannot jump very far. Contact is broken either by springback as the pressing load is released or by the absence of an adequate number of fines in the composition. When there is an inadequate number of fines, the volume decreases as the matrix shifts, leaving voids around the grog pack. In order too avoid this, the grog pack must be a little loose so that it can be carried down with the shrinkage of the matrix. In this particular case, there are not enough fines to go around because they have agglomerated and are not functional in the shrinkage structure. Mixing is important.

Sample Preparation

This section provides some helpful tips rather than giving detailed instructions on how to make sections. Detailed procedures are specific to individual materials and cannot be generalized except in an overall manner. The best source of detailed information is from colleagues working with the same materials who have previously worked out the methods by trial and error or from knowledgeable suppliers.

Polished sections. The first step is to select the location and orientation of the place to make the section. This could be an edge where a reaction has taken place or an orientation to see if the grains are aligned. Smaller is better unless there is a reason to make a large section. It is much easier to polish a small cross-section than a large one; 1 by 1 cm is a convenient size.

Sawing. When starting with an irregular shape or a large sample, sawing is the next step. Use a continuous-metal-rim, thin, diamond blade. Sawing often results in deep damage to the surface when severe, making it advisable to use a fine diamond grit (about 100 grit or finer), a low blade surface speed (1000 inches/min.), and a slow feed rate, sometimes regulated by a weight pressing the blade against the sample. Use water as the coolant, sometimes with a cutting fluid additive. Two figures of diamond saws are included. Figure 11.37 is of a small, slow-cutting saw that is useful for small samples.

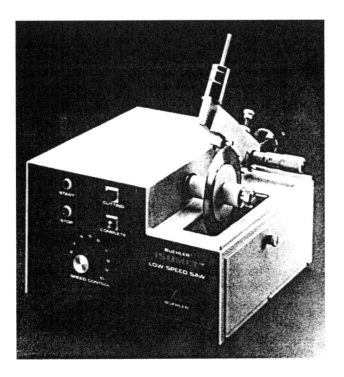

Figure 11.37: Small Diamond Cut Off Saw. Useful for light laboratory work; it produces a smooth cut. (Courtesy of Buehler)

Cutting is slow, but there are other things that can be done in the waiting period. As the cut is easy and slow with a fine diamond blade, there is relatively little damage done to the sample surface. By contrast, a rough cut will shatter the surface to a considerable depth, requiring a lot of lapping to remove it. A continuing problem is holding an irregular shape in the saw vise. If necessary, the sample can be mounted in a polymer cast to make a regular shape suitable for gripping in the vise. Another alternative is to rough-saw a regular shape and then make the section cut. Figure 11.38 is of a larger saw for more heavy duty work that is still of a lab scale.

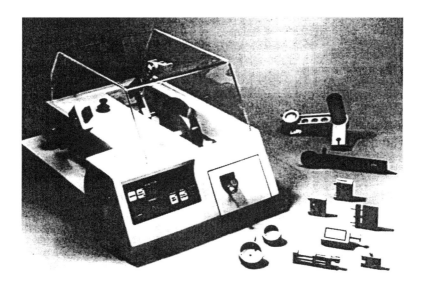

Figure 11.38: Laboratory Sized Diamond Cut Off Saw. Used for general laboratory work. (Courtesy of Buehler)

Impregnation. When the sample is porous, it is impregnated with a polymer. Epoxy is most commonly used. Refer back to Figure 11.17 of an apparatus used for saturating a porous sample for specific gravity determinations. The same apparatus is used for infiltrating the sample with the epoxy polymer. The procedure is the same. A vacuum is pulled, the epoxy is let in, and the chamber is pressurized to about 60 psi. By using pressure, a higher viscosity epoxy can be used. The epoxy will have a higher molecular weight, less shrinkage, and better physical properties. The resin will spatter as it is admitted into the chamber, however that will not be a problem as long as the tube is far enough below the top of the beaker.

After use, the apparatus must be cleaned. The valve is a stopcock-type dismountable by a snap ring, making it easy to clean off the residual resin with a solvent such as acetone. Chaining the snap-ring pliers to the apparatus ensures that they will always be available.

Use a casting resin to mount the sample, as hot molding phenolic overheats the epoxy.

Another procedure for infiltration is to use a glass frit with a low melting point. Melt it on the surface over a burner. Lead borosilicate glasses melt as low as 450-550 °C, low enough to prevent reaction with almost all ceramics. Infiltration is shallow as no pressure assistance is used, making it necessary to limit the amount of surface that can be ground off. Frit is advantageous for having a higher modulus than a polymer, providing greater support for the sample at the edges of the porosity. When the porosity is very fine, frit infiltration is the preferred technique and may be the only option for making the section.

Premolding Treatment. Scratches are usually caused by pieces chipping off the edges of the section and being dragged across the surface. A few scratches do not make any difference, but too many scratches obscures much of the surface. Grinding a small bevel on the top edges of the sample greatly reduces the tendency for the edge to chip. Lapping off the excess epoxy or frit prior to mounting helps the sample to lie flat during mounting.

Mounting. If the sample is to be polished by hand, there is no need for mounting. When the sample is to be polished in an apparatus, a mount must be made for it to geometrically fit into the polisher. Polishing equipment is the preferred method, especially when there is more than one sample. Offhand polishing is quick and dirty and often used for a quick look. Samples can be mounted in a phenolic resin by hot molding or cast in a mold with a polyester resin. Some polymers have a strong exotherm and will char and bubble during curing. Large castings are susceptible to overheating. Procedures for mounting in a phenolic resin, as given by the vendor, do not allow enough time for the resin to crosslink. Do not follow the manufacturing directions in this case, but give the molding additional time (+5 min.) to crosslink at the molding temperature. Otherwise, since the phenolic is soluble in organic solvents, it will produce a mess when the polishing oil is cleaned off with a solvent.

Lapping. Sawing leaves a rough surface, with damage extending into the interior of the sample. This must be lapped off with abrasives using water as a cutting fluid. Figure 11.39 shows a set of laps mounted in a polishing bench.

Figure 11.39: Polishing Lapidary Table. Can be used for wet grinding and polishing if kept clean. (Courtesy of Buehler)

The photo shows a setup for polishing. But, the bench can be used for both lapping and polishing. When the ceramic sample is hard, it is necessary to use diamond laps where the grit is held on the surface with nickel plating. These laps are thin and are conveniently held in place with a magnetic disc. This makes it easy to change grit sizes. Grit sizes start for many applications at 60 grit and then go sequentially to about 100 grit and finally 220 grit diamond. Finer sizes do not work well with this system as the lap surface seems to glaze over and ceases to cut. All lap surfaces should have voids. For loose abrasives, these voids are a spiral groove machined into the lap surface. The nickel-plated laps are best when perforated with an array of small holes. These are seen in a sketch of the interface between the sample and the lap surface, as depicted in Figure 11.40.

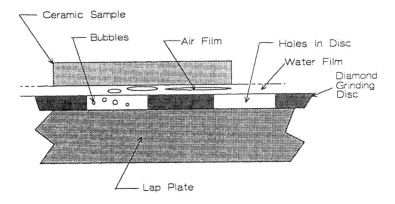

Figure 11.40: Nickel bonded Diamond Lap with Perforations in Surface. Fastens with a magnetic disc for easy removal.

Voids on the lap surface help to break the "suction" between the sample and the lap surface and supply lubricant (water) to the interface. The sketch shows air bubbles that are drawn out into films on the interface. Air, as a gas, expands when pressure is lowered. This expansion allows the sample to glide freely over the lap surface. A solid lap surface "grabs" the sample, seemingly pulling it from one's hands. If the lap runs dry, the sample overheats, and there is no means to carry away the swarf. Laps made from ceramic materials are now available and remain flat longer. On these, a diamond slurry of 9 to 30 μm diamond grit is used.

Loose abrasives can also be used for rough lapping when the ceramic is soft enough. A drip of water keeps the lap lubricated. Silicon carbide from 80-600 grit is a common abrasive for these ceramics. The lap wheel is cast iron with a spiral groove. These wear and require frequent re-machining. For this kind of installation, a sump is necessary to prevent clogging of the drains. Also, pitch the drain to the sump steeply or it will clog. Alumina powder is sometimes used for fine lapping in the range of 600-1200 grit. Remember to keep everything clean.

Polishing. Fabric-covered laps are used for polishing with diamond-polishing compounds in a lubricant liquid. Eight inch diameter laps are most common. Polishing machines are available and one type is shown in Figure 11.41.

Figure 11.41: Polishing Apparatus. A variety of polishing equipment is available. This is a common type. (Courtesy of Buehler)

The machine can hold five or six samples. When there is an imbalance, a dummy section can be inserted in a blank opening. Newer machines do not require this. Fabric choice is important. Generally, a low nap fabric is used that can be nylon, a bonded-felted material, or cotton. Alumina polishes are best on nylon, while zirconia polishes are best on the bonded-felted material. Glasses polish on stiff foamed polyurethane with CeO_2 polishing compounds. Plastics use alumina polishing compounds on higher nap cloths. This type of cloth is also useful for making a relief polish, which is where the softer materials are removed to a greater depth than the harder ones. This brings out the difference between the two phases, allowing their easy recognition. However, in general, a flat polish is sought with the features brought out by etching. Very flat surfaces can be produced on lead laps, but require expertise. The latter method is not common in the ceramics lab.

Polishing compounds are carefully-sized diamond powders that range from 30 to 1/4 μm. Polishing proceeds progressively from the coarsest to the most fine size. Clean thoroughly between abrasive sizes. An ultrasonic bath with a detergent or solvent is useful. Reflective surfaces develop with the 9 μm abrasive. When the surface is no longer reflective, change the procedure, usually with a different cloth. Final polishing is with 3 μm diamond for most purposes, but 1 and 1/2 μm abrasives are often used. If the difference is not discernable, then the extra steps are not especially worthwhile. During polishing, the rough-lapped surface is gradually removed. That surface contains scratches that have uneven scratch roots. These can be differentiated as they line up on the partially-polished surface, whereas porosity does not. Figure 11.42 is the polished surface of a dense alumina sample.

Much of what seems to be porosity is not. Rather, when the dark spots line up, they are scratch roots. Also, pores are often round where root pits are not. Scratch pits require additional polishing to remove them. However, it is often less work to go back and re-lap with a fine abrasive and then start to polish all over again. It depends on the depth of the pits.

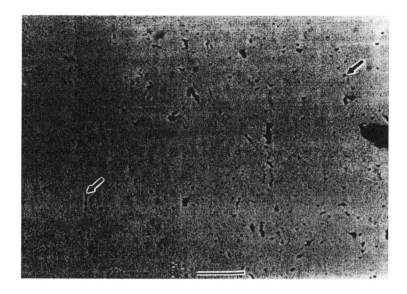

Figure 11.42: Dense Fine Grained Alumina Ceramic. Polished section, plane light. When "pores" line up they are scratch roots, as shown by the arrows. Scale bar 200 μm.

Etching. Etching is often done to delineate microstructural features such as grain boundaries. Fine-grained, dense ceramics are usually thermally etched at about 1200-1300 °C for 15-30 minutes, presuming that the polished sample was first removed from the plastic mount. This is a preferred technique for alumina and zirconia. When the grain boundary phase or matrix is a glass, etching can be done with HF vapor by holding the sample with tongs above the acid. In other cases, the sample is immersed in the acid, which is diluted to produce the right amount of etching. A third way to etch a sample is to place it in a molten flux for a brief period of time. Alkali fluorides are common as fluxes, and the process can be done over a gas burner.

The top surface of the polished section must be parallel to the top surface of the stage for the section to remain in focus. To accomplish this, place a thumbnail-sized piece of modeling clay on a glass slide and level it with a leveling press, which can be obtained from an accessories supplier.

Thin Sections. Thin sections are viewed with transmitted light, either plane or polarized. There are many techniques determining optical properties, but only a few are applicable to ceramics. These few include, when there is an advantage to viewing the material in depth through a cross-section, birefringence, microstructure, orientation, and details such as those on grain boundaries. A prerequisite is that the grain size is larger than the section thickness, in the range of 30-50 μm.

Making thin sections is difficult, especially for hard and tough ceramics. Residual stresses are the biggest concern for the following reasons. A section is made by slicing a thin piece of the ceramic, lapping one side flat, and cementing it to a glass slide. Epoxy adhesives are common. Next, the section is ground down on the opposite side to the required thickness and covered with a cover glass. Grinding or lapping result in residual stresses in the surface.[2] Stress levels are affected by the coarseness of the abrasive; coarse abrasives impart greater residual stresses than fine abrasives. The side of the section against the slide was finish-lapped with a fine abrasive to give it a smooth finish. The other side is lapped initially with a coarse abrasive producing a difference in stress between the two surfaces. That the section tends to curve has been observed where the difference in calculated stress is as high as 30,000 psi. Spalling then occurs on the edges of the section, with the fragments tearing out additional material. If fortunate, the center of the section might be preserved.

High index of refraction ceramics such as zirconia must be polished on both surfaces because light scattering will cause foggy viewing. This situation is even worse, as the first polished surface is relatively stress-free while the other surface is being lapped with coarse abrasives and is in a state of high stress. Also, the first surface is smooth, with little surface area for the cement to grip. Spalling is severe under these conditions.

What can be done to reduce spalling? There are a few experimental solutions and some that are incomplete.
- Roughen the glass slide by lapping with 600 grit SiC.
- Use a stronger adhesive, hot melt?
- Lap the observation side of the sample with fine abrasives only.
- Use gentle and slow lapping, with a polishing apparatus.

Other problems with making thin sections include keeping the section flat, lapping a wedge, forming a dome, slicing finger tips on the slide edges, and grinding off finger tips. Keeping the section flat and parallel takes practice and concentration. There is a natural tendency to grind a dome as the slide rocks on the lap. It helps to grind preferentially on the center of the dome by placing a fingertip and pressing down directly on the high part.

With enough practice, the slide can be held flat. Standing at the lap is the preferred position, provided that the table is of a suitable height. The slide can be held in a special holder as shown in Figure 11.43.

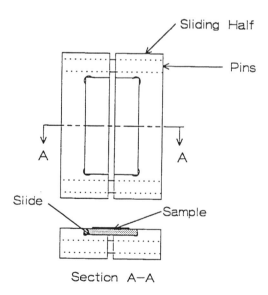

Section A-A

Figure 11.43: Thin section Slide Holder.

There is both a feel and observation when the slide is held flat. Body motion is much like a dance, where the wrists and arms are held stationary and the body rotates and shifts from the hips traversing the section across the lap surface. Periodically, turn the slide over and look to observe a wedge, with one end or side thicker than the other. The trailing edge of the slide laps faster than the leading edge because of the torque on the slide. Simply turn the slide around so that the thick part of the wedge is on the trailing edge. It also helps to press down harder on the thick side. Alternatively, the slide can be tipped up to hold it parallel to the lap surface. Then, look at it often and make further corrections.

As for cutting injury to the fingers, dull the slide edges slightly and use a slide holder such as the one shown in Figure 11.43. However, the best feel is when the slide is held directly with the fingers. Proceed with caution since the combination of vibration and cold water prevents pain in the fingers, often rendering an injury as a surprise. With practice and healing, thin sections will be made flat and of the right thickness. After much practice, the craft is mastered.

When using loose abrasives, the lap will wear unevenly, with the edge and center wearing less than the middle annulus. Because of the lap curvature, the lap will grind a dome on the section. Periodically, the lap must be machined flat. These laps have a spiral groove re-machined into the surface for distributing the coolant and abrasive, and breaking the suction.

Metal-bonded diamond laps dull and cease to cut. They can be rejuvenated by grinding a gritty ceramic such as an insulating fire brick against the surface. This wears away the bond and brings up additional diamond asperities. The procedure does, of course, wear the lap.

Care of the Optics. Microscopes have soft glass lenses that scratch easily and corrode in acidic gases. To avoid touching the objective lens to the sample, always lower the lens to where it is just above the sample surface and then raise it to focus. This becomes more difficult at high magnifications where this distance is small. When looking at granular materials on a slide, always use a cover glass and an index of refraction liquid. Using either a cover glass slide or a polished section ensures that the lens will not be damaged if it inadvertently touches the sample as it is in contact with a flat surface.

Keeping the lenses clean helps to obtain a sharp image. Clean properly to avoid scratching the lenses. Avoid rubbing an abrasive particle across the lens surface. Blow the lens off with a canned gas, or brush it lightly with a soft camels hair brush, then clean it with an organic solvent such as acetone and lens paper. Avoid vigorous rubbing. Facial oils are the main ocular contaminants. These oils pick up grit that can damage the lens if it is rubbed. Most of the oil is around the rim of the lens and can be cleaned using a Q-Tip moistened with acetone. Rubbing the bulk of the lens only spreads the oils out over the surface, so it is necessary to clean the edges first. Cleaning the edges first with a detergent works well on eye glasses. Objective lenses remain cleaner as they are out of the way. Cleaning is done by the same procedure, except when the dirt is abrasive, such as being splashed with a slip. Be very careful not to rub the lenses until all of the slip is blown or lightly brushed away.

Check List, Optical Microscopy

- Stereo-Binocular
    - 40X max.
    - Oblique illumination
    - Flexible illuminator
    - Focus eye piece
- Universal
    - 1000X max.
    - Polished surfaces
        - Vertical illuminator adjustment
        - Measure: grain intercept, % phases, shape
        - Oblique illumination, 100X max.
    - Thin sections
        - Loose grain
        - Optical contrast
        - Birefingence
        - Accessories: filar, rotating stage, mechanical stage
        - Measure: grain intercept, percent phases, shape, orientation, stresses

- Sample preparation
    - Polished sections
        - Sawing
        - Impregnating: polymer, frit
        - Ease edges
        - Mounting: casting, molding
        - Lapping
        - Polishing: diamond abrasive, cloth, medium
        - Etching: thermal, acid, flux
- Thin sections
    - Residual stresses
    - Plucking problems
    - Flatness
    - Loose lapping abrasives, cast iron laps
- Care of the optics
    - Grit removal
    - Solvents
    - Cover glass

Electron Microscopy

Described below are Scanning Electron Microscopes (SEM) and Transmission Electron Microscopes (TEM). Both microscope types can be equipped with chemical analytical capability. There are two types of analytical capability, energy dispersive (EDX) and wave length dispersive (WDX). Both are ways of analyzing the X-ray fluorescent spectra emitted from a spot on the sample surface energized by the electron beam.

Electron microscopes use an electron beam rather than visible light to probe the sample. An electron beam is more energetic and capable of resolving to much higher magnifications. The beam is focused with electromagnetic lenses to produce an image on a monitor or photographic film.

Scanning Electron Microscopes (SEM)

Scanning electron microscopes scan across the sample surface with an electron beam. They are useful over a wide range of magnifications from a few diameters up to about 20,000X. An instrument is shown in Figure 11.44.

Figure 11.44: SEM Instrument. There are several good manufactures of SEM equipment. (Courtesy Amray, Ferro)

Modern instruments, more user-friendly than those of 10 years ago, continue to improve. Software is extensive and versatile. While capable of higher magnifications, surface charging occurs and blurs the image since most ceramics are electrically insulating. Scanning the sample surface results in a great depth of field. This capability is very useful as the entire vertical dimension of the sample surface is in focus, as will be seen in the subsequent discussion. A very common application of the SEM is to observe microstructure when the grain size is too small for an optical microscopes. Figure 11.45 is the thermally-etched, polished surface of a TZP sample.

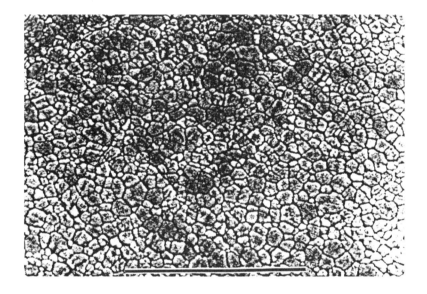

Figure 11.45:   Yttria stabilized Zirconia(TZP) Fine-Grained, Dense Ceramic. Thermal etch. SEM Photograph. Scale bar 10 μm.

The ceramic is fully dense with an average grain intercept of less than 1 μm; evidence that it was well-crafted.

Another common use of a SEM is to look at fracture surfaces in order to identify the fracture origin. Figure 11.46 depicts the fracture surface of a TZP ceramic with a rather evident fracture origin.

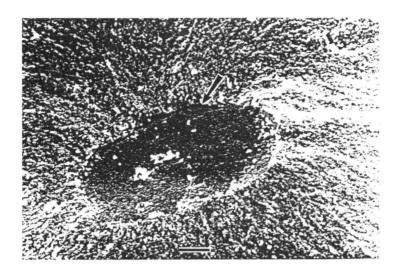

Figure 11.46: Microstructural Flaw in a Yttria-Stabilized Zirconia Fine-Grained, Dense Ceramic. Fracture surface. SEM Photograph. The arrow points to what appears to be an agglomerate. Scale bar 10 μm.

The arrow points to the flaw that appears to be an agglomerate. If it were a different material, it would probably display cracks around the edges as the firing shrinkage would likely be different. With the analytical capability of EDX, the composition of the flaw can be determined, with the intent of removing this material from the powder so that it won't happen again. Observing fracture origins is common, as the ceramic selectively chooses to break there.

What else can be learned from looking at SEM photos of ceramics? Take, for example, the surface shown in Figure 11.47.

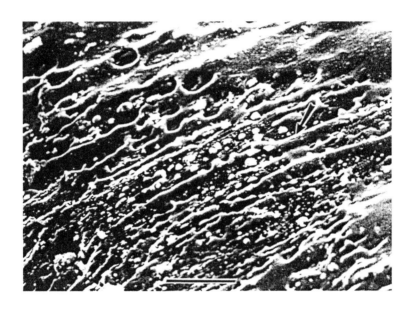

Figure 11.47: Wear Surface on an Alumina/Zirconia Abrasive Grain. SEM photograph. The stainless steel phase was molten during machining. Scale bar 10 μm.

Ceramic materials are used as abrasives; this one is an alumina-zirconia-fused grain used for machining stainless steel. It is obvious from the photo that the temperature was above 1429 °C, which is the melting point of 304 stainless steel.

It is written that polishing is a process of finer and finer abrasion. This is not always true, as seen in Figure 11.48.

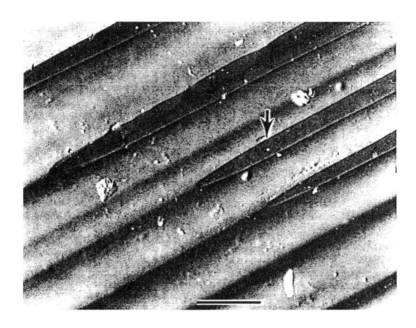

Figure 11.48: Wear Surface on a B$_4$C Colloid Mill Plate. SEM photograph. The wear surface is grooved and polished. The arrow points to a groove with residual alumina. Scale bar 200 μm.

     This is the wear surface of a B$_4$C wear plate in a colloid mill used for comminuting an alumina powder. Some of the alumina is seen stuck in the wear grooves on the surface. Note two things. First, boron carbide is much harder than alumina and cannot be abraded by it. Second, the wear surface is polished, implying that wear is by a molecular-scale process where abrasion is by a relatively macro-scale process. Polished wear surfaces are usually the result of diffusion and chemical processes, rather than mechanical. Thinking of wear as centrally related to mass transport reduces confusion. There are exactly four mass transport processes:
- diffusion,
- viscous flow,

- fracture and kinetic displacement, and
- plastic deformation.

Examination of wear surfaces often indicates which of these four, alone or in combination, is the wear mechanism. This often leads to conclusions as to what can be done to lower the wear rate. In the case of the B_4C wear surface, the process was likely chemical, involving diffusion and viscous flow. A change of material is indicated.

The next example of the power of the SEM is concerned with the heavy-duty machining of steel.[3] Steel has a structure where the carbon-containing phase is clustered as Fe_3C ceramic domains, seen in Figure 11.49.

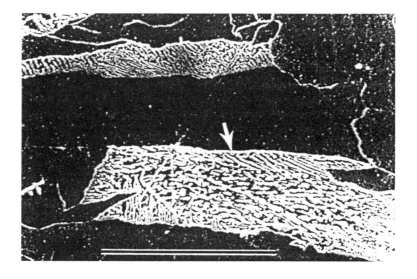

Figure 11.49: Microstructure of Steel. SEM photograph of a polished section. The arrow points to the Fe_3C pearlite ceramic phase. The matrix is iron. Scale bar 10 μm.

There are two phases present, alpha iron and pearlite (Fe_3C). Iron is a malleable metal. Pearlite is a brittle, hard ceramic. When the surface is severely deformed, the pearlite clusters mash down into layers as seen in Figure 11.50.

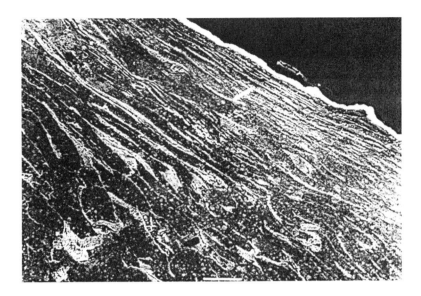

Figure 11.50: Layers of Flattened Pearlite due to Shear caused by Rough Machining. SEM photograph of a polished section. Arrow points to a flattened pearlite phase. Scale bar 10 μm.

The shear surface is at the top of the picture and, in that region, the layers of pearlite are flattened. Using the SEM at higher magnification on this region near the surface, one can observe a high concentration of Fe_3C platlets to the point that the surface is nearly all pearlite, as seen in Figure 11.51. (Incidentally, carbon was detected in this phase with EDX.)

Figure 11.51: Concentration of the Pearlite Ceramic on the Wear Surface as shown by the arrow. SEM photograph of a polished section. The wear surface is almost all pearlite. Scale bar 10 μm.

From these observations, one can deduce the mechanism. Start with an analogy. Take a thin sheet of plywood (pearlite) and cover it with an inch of Jello (alpha iron). Build that order up for several layers. Now, jump on the stack. Obviously, the Jello will squirt out on the sides leaving just the plywood on the surface. That also occurs when machining the steel. This understanding is made possible from contemplation of the SEM.

As mentioned earlier, an SEM can be equipped with analytical capability, either energy-dispersive or wavelength-dispersive techniques. Using these techniques, the chemical composition of the surface can be determined. As with all analytical methods, there are some limitations, but these are not especially restrictive and the analytical tools are generally very useful. The combination of a wide range of magnification, great depth of field, and analytical capability makes the SEM one of the best problem solvers around.

Energy Dispersive Analysis. In this method, the secondary fluorescent X-rays are separated into a spectrum by their energy. High frequencies have more energy than low frequencies and are spread out into a spectrum of peaks of varying magnitude. The instrument is smart and will label each peak as to what element it represents. As there is some overlap, the instrument will present a few choices, most of which are improbable, as they are rare in ceramic formulations. Elements from B (atomic number 5) on up, with a few exceptions, can be analyzed. Figure 11.52 shows an EDX attachment for either an SEM or TEM.

Figure 11.52: Energy Dispersive X-ray (EDX) Attachment for an Electron Microscope. Measures chemical composition of a surface. Very useful technique. Virtually mandatory. (Courtesy Oxford)

The figure is of the sensor coupled to electronics not shown here. At low atomic numbers, sensitivity is low as the X-rays are soft (i.e., weak and easily absorbed). The same problems exist with energy-dispersive analysis as with bulk X-ray fluorescent analysis, since they are similar methods. The main problem is absorption by the matrix, which is more acute when the matrix contains elements that are strongly absorbing. Matrices with light elements are less troublesome. A set of related standards makes the method quantitative, with perhaps two significant figures.

Another useful ability of a SEM is the selectivity of the area to be analyzed. This can range from a 1 μm spot to a broad field. While the electron beam is about 1 μm in diameter or less, the radiation spills over and penetrates into the sample, limiting the smallest field to a few cubic micrometers. An additional attribute of this method is the selectivity it provides for the analysis. For example, the surface can be mapped for a single element, showing its distribution and rough concentration.

We will now discuss a ceramic problem where extensive wear of alumina abrasives occurs when machining titanium. Three figures are used to illustrate the example. Figure 11.53 shows a groove cut in titanium by an alumina cutter.

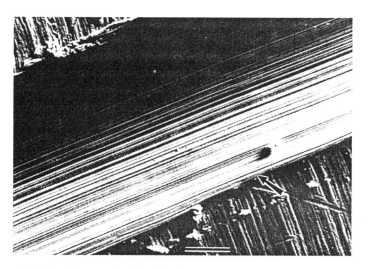

Figure 11.53: Groove Cut in Titanium with an Alumina Abrasive Point, SEM Photograph. The groove is striated. Scale bar 100 μm.

As can be seen, the groove is striated and smooth along the direction of the cut. Figure 11.54 is of the same field, except that the image was made from just the Al spectral lines.

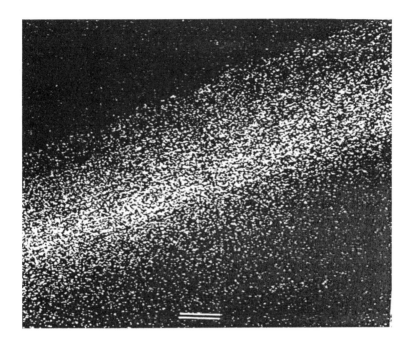

Figure 11.54: Aluminum Concentration in the Same Groove (as in previous figure). SEM photograph with EDX Al Spectra. Aluminum has been transferred to the titanium metal. Scale bar 100 μm.

Aluminum (Al) is being transferred from the abrasive tip to the titanium (Ti) surface, without any visual evidence of particulates. The transfer is not by reduction of Al_2O_3 by Ti as the thermodynamics does not allow this to occur. If direct diffusion of Al into the titanium is not occuring, then something else is. Consider that Al and Ti are not the only phases present. There is also O_2 present in the system, which is responsible for the material transfer in that Ti is oxidized at the high surface temperature and an aluminum titanate compound is then formed as an interstitial film. A traverse can be made across the groove using only the aluminum spectra, as seen in Figure 11.55.

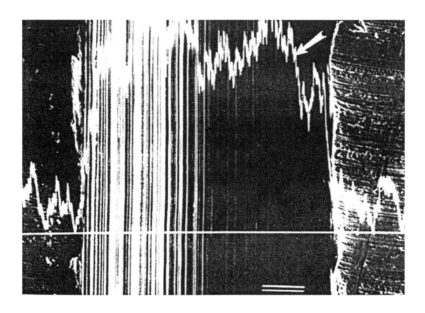

Figure 11.55: EDX Traverse across the Same Wear Groove, SEM photograph. Scan shows the distribution of aluminum. Arrow points to maximum Al. Scale bar 100 μm.

In the line scan, the concentration of Al is proportional to the height above the base line. When the curvature of the groove is considered, the Al distribution is fairly flat across the traverse. Line scans by the SEM and EDX are useful tools for studying surfaces, in particular when reactions are occurring. Wear of the alumina abrasive, in this case, was by diffusion, after oxidation of the Ti surface.

Entertain one more example of how the SEM and EDX can be used to unravel ceramic puzzles using the following four figures. Firstly, examine Figure 11.56.

Figure 11.56: Wear Surface on an Al_2O_3/ZrO_2 Abrasive Grain after Machining Titanium. SEM photograph. Surface shows a reticulated pattern of attached metal. Scale bar 10 μm.

This is the wear surface of an Al_2O_3-ZrO_2 abrasive grain after it was used to machine Ti metal. It is evident that the Ti was molten because the material has formed rounded shapes indicating a whopping 1660 °C on the surface.

Figure 11.57 is the Ti distribution map, showing its position on the same field of view.

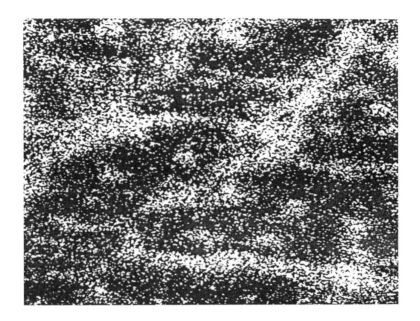

Figure 11.57: Titanium Distribution on the Same Wear Surface. SEM photograph with titanium radiation. Titanium follows the reticulated pattern. Scale bar 100 μm.

As would be expected, the location of the Ti and that of the previously-molten metal are coincident. On the same field, Figure 11.58 shows the distribution of Zr.

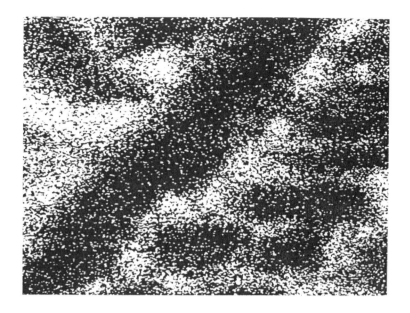

Figure 11.58: Zirconium Distribution on the Same Wear Surface. SEM photograph with Zr radiation. Zirconium follows the same pattern as titanium. Scale bar 100 μm.

Figure 11.59 is of the Al distribution on the same field.

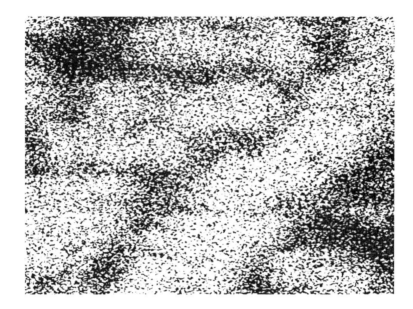

Figure 11.59: Aluminum Distribution on the Same Wear Surface. SEM photograph with aluminum radiation. Aluminum is not wet with the titanium. Scale bar 100 μm.

Here are the following conclusions.
- Ti was molten on the surface as it was machined.
- The distribution of the Ti was localized.
- Zr distribution follows the same pattern as Ti.
- Al distribution does not follow the same pattern.
- Ti wets the ZrO_2 phase on the abrasive surface.
- Wear probably occurs by a reaction between the Ti and ZrO_2 in the presence of O_2, where the mechanism is potentially viscous flow of the liquid titanium.

Energy dispersive analyses are fast and easy to obtain, depending on the precision needed and the concentration of the element being analyzed.

When dealing with low concentrations, a peak is selected and a count made on that location. Then, as with bulk X-ray analyses, the adjacent background is counted and subtracted.

Wavelength Dispersive Analysis. The same fluorescent spectra taken from the sample surface can be separated by X-ray diffraction. To do so, the X-rays are collimated and impinge on a diffracting crystal selected for its lattice spacing. To cover the range of elements, it is necessary to include three crystals with different lattice spacings. The crystal diffracts radiation and a goniometer scans the spectrum just as in X-ray diffraction. The scan takes time to complete and thus is a slower process than energy-dispersive analysis. When the amount of a single element is sought, the angle is set on the peak and a count taken. The intensity is proportional to the amount present and is subject to the same restrictions as for other fluorescent analyses. While a slower process that requires more work, wavelength-dispersive analysis is capable of greater precision and sensitivity than energy-dispersive analysis.

Transmission Electron Microscopy (TEM)

TEM is a method where the electron beam is transmitted through the sample rather than just impinging upon its surface. The range of magnification is very broad, from a few hundred up to 1,000,000X with some instruments. Lattice planes of heavy elements can be resolved as can individual dislocations. The TEM can also produce an X-ray diffraction pattern that yields crystallographic information. In ceramic work, the TEM is used mostly for viewing powders, surface replicas, and thin sections. Figure 11.60 shows a TEM.

The instrument is essentially a column with an electron source, electromagnetic lenses for focusing the beam, a high potential for accelerating the electrons, and a stage for holding and manipulating the sample. Below the sample, there are sensors for collecting the image. The entire column is evacuated to a very low pressure.

Limiting is that the electron beam is not penetrative, so the sample must be very thin (<1 μm) or have low absorption and also be thin. There are techniques for looking at materials: dispersed powders that are viewed in silhouette on a carbon film, a carbon film replica that is shadowed with a heavier element to show surface detail, or a sample that has been thinned down to where it is transparent using specialized techniques.

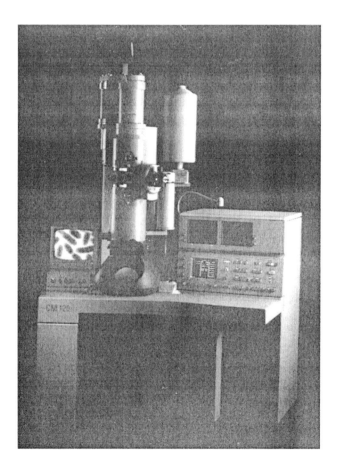

Figure 11.60: TEM Instrument. Several good instruments are available with many attachments. (Courtesy of Philips)

Dispersed Powders. A carbon film supported on a wire mesh is coated with a dispersion of the powder. The electron beam can penetrate the carbon, but not the particles that are seen only in silhouette. Figure 11.61 shows SiC particles on a carbon substrate.

Figure 11.61: Silicon Carbide Particles viewed with the TEM. A dispersion of particles on a transparent substrate can be directly viewed with a TEM. Scale bar 1.0 μm.

When the particles are fine enough, they are transparent to the electron beam and additional information can be obtained on their structure. Depending on the absorption of the particular material, the limit on size for transparency is about 1/2 μm.

Surface Replicas. In this technique, a surface is coated with a stripping film that is pealed off after it dries. It is then placed on a mesh for support and vacuum-coated at an angle with a heavy element, such as Au or Pd. Figure 11.62 is a surface replica of an alumina-cutting tool wear surface, in a stereo pair.[4]

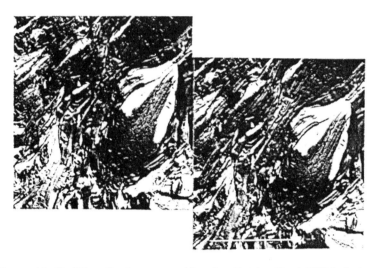

Figure 11.62: Wear Surface on an Alumina Cutting Tool. TEM photograph of a replica film shown as a stereo pair. Stereo gives a three dimensional view of surfaces. Scale bar 100 μm.

The wear surface is striated and contains ellipsoidal bumps on the surface (when using replicas, the vertical dimension is reversed and protrusions appear as depressions). Wear occurs by plastic deformation of the Al_2O_3-Fe_2O_3 phase on the surface by reaction of the alumina tool with steel in the presence of oxygen. This phase is a spinel that has cubic symmetry and can deform in bulk by plastic deformation when the temperature is high and the material is restrained by pressure from fracturing.

This techinque can be taken to the field and applied to large structures. Compared with a direct view as with the SEM, the TEM replica technique provides much more surface detail. Stereo pairs are useful when the surface has relief.

Another replica technique involves coating the sample with carbon and subsequently dissolving it. This leaves just the replica film that is then vacuum-coated and observed in the TEM, as seen in Figure 11.63.

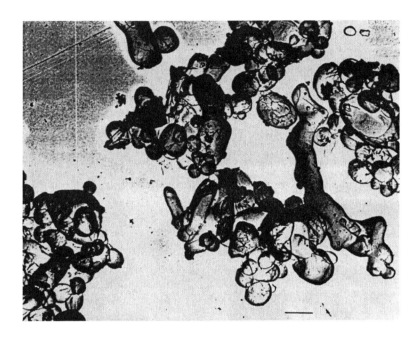

Figure 11.63: Zirconia Particles, Replica viewed with a TEM. Surface detail is exquisite with TEM. Scale bar 1.0 μm.

The subject is monoclinic zirconia, which was produced by cracking zircon with lime at a high temperature. Surface detail is exquisite and far better than obtainable by other methods.

TEM can also be used to look at microstructure, as seen in Figure 11.64.

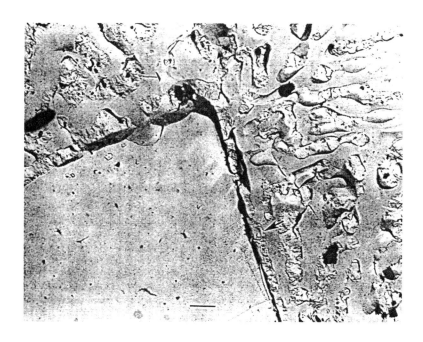

Figure 11.64: Fused Cast Al_2O_3/ZrO_2 Fracture Surface. The large flat area is an alumina crystal. The arrow points to a zirconia film encasing the crystal. Matrix is solidified alumina/zirconia eutectic. Scale bar 1.0 μm.

The sample is an arc-fused mixture of alumina and zirconia. This view is a fracture surface from which the sequence of crystallization can be deduced. There are three distinct regions in the structure. The large, flat surface on the lower left is an alumina crystal that was the first phase to crystallize out of the melt. As the melt composition became more concentrated with zirconia, it became supersaturated and plated out on the surface of the alumina crystal as a thin film about 1/2 μm thick. The alumina crystal is then isolated from the melt, which then crystallizes as an eutectic mix of both alumina and zirconia. It is this eutectic structure that imparts toughness to the material.

Basic research institutes and universities are science-driven, while industry is technology-driven. The TEM is a valuable tool for scientific studies on ceramic materials. For this, thin sections are often used where grain boundary structure, some aspects of crystal structure, and dislocations can be observed. The section must be less than 1 μm thick. These are made with a dimpler that abrades a hole through the normal petrographic thin section. The edges around the hole are thin enough to view. One can also dissolve the center of the dimple to reach the necessary thickness. While these techniques are not normally used in industrial labs, they are useful when employed. Figure 11.65 shows a TEM thin section. The sample is MgO-partially-stabilized zirconia showing tetragonal precipitates in the structure.[5]

Figure 11.65: Tetragonal Zirconia Inclusions in Partially MgO-Stabilized Zirconia, TEM Thin Film. Elliptical inclusions of the tetragonal phase are seen at the arrow. Scale bar 200 nm.

This structure is centrally vital to the toughening of zirconia ceramics. Other studies by the same group using TEM techniques conclusively proved that the transformation is indeed martensitic in zirconia.

Scanning TEMs are also available allowing a view in greater depth, but bear in mind that the sample is very thin and limited depth is available.

Like the SEM, the TEM can be equipped with EDX and/or WDX analytical capability. Obviously, this is of no help when using a replica, as the sample is not present. However, it is valuable on thin sections, especially when working on very fine structures such as grain boundary phases.

Check List, Electron Microscopy

- SEM
    - 20,000X
    - Great depth of field
    - Views surfaces
    - Energy dispersive analysis
        - Quick and easy
        - Composition maps
        - Composition traverses
        - Selective areas
    - Wavelength dispersive analysis
        - Slow scanning
        - Precise
        - Trace elements
        - Count on a peak to increase sensitivity
- TEM
    - 1,000,000X
    - Electron diffraction
    - Powders
        - Silhouette
        - C film + dissolution
    - Replicas
        - Shadowing

Thin sections
Direct observation
Grain boundary detail
Dislocations

5.0 PHYSICAL PROPERTIES

What has preceded this section leads up to developing the physical properties of the ceramic that are needed to fill an application. This section is concerned with how these measurements are made and discusses eight properties: modulus of rupture (MOR), tensile strength, compressive strength, modulus of elasticity (MOE), hardness, fracture toughness, wear resistance, and thermal shock.

Modulus of Rupture (MOR)

MOR tests are conducted on a test machine where stress is measured as the sample bends until it fractures. Figure 11.66 depicts a lab test machine.

A test machine is essentially a rigid frame with a moving head that imposes stress onto the sample. The machine is very carefully designed and constructed so that movements are smooth and even. Force is measured with a load cell and displacement with an extensometer. Alternatively, deflection can be measured as cross head motion, after subtracting out machine compliance. As seen in the figure, the instrument is computer-controlled. The test machine operates at different speeds that are fairly slow for ceramics: about 0.01"/min. Ceramic grain size influences the size of the test bar. A coarse-grained refractory requires a much larger test bar than a 1 μm grained-dense ceramic. Typically, 1/2"x 1"x 7" test bars broken on a 6" outer span are appropriate for the coarse-grained refractory. For a fine-grained ceramic Mil.Std. 1942 is useful. These test bars are specified in a few sizes, with the 3 x 4 x 45mm bar with chamfered edges most common. This size is tested in four point bending with a 20 mm inner span and a 40 mm outer span. The surfaces must be finely ground and the edges chamfered. Well-thought-out fixtures are commercially available with allowances for alignment and bar flexure as it bends. A four-point-bend fixture is seen in Figure 11.67.

Figure 11.66: Physical Test Machine. Used for strength and fracture toughness testing. Many attachments are available. (Courtesy of MTS)

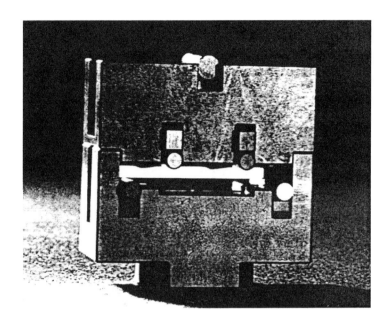

Figure 11.67: Four-Point-Bend Fixture. Used for MOR measurements. (Courtesy of MTS)

As the sample bends, the length between the loading lines changes, becoming shorter on the top and longer on the bottom. Clearance is provided so that the loading pins can roll as these distances change. Fixtures also provide side-to-side flexibility so that the test bar and loading lines comply with one another. These ceramics are brittle materials with a high modulus of elasticity, so they are not forgiving of uneven loading or misalignment. Four-point bending is preferrable to single point as a larger volume of the sample is subjected to the maximum stress. Since fractures in brittle materials are initiated by flaw origins, the chances of having one in the stress zone is higher, and the results are more representative.

Attachments for this type of apparatus are expensive and can quickly impoverish a lab's budget. One attachment that may be needed is the capability of measuring MOR at elevated temperatures. An attachment for this purpose is shown in Figure 11.68.

Figure 11.68: High Temperature Bend Test Apparatus. (Courtesy MTS)

Data on MOR is almost always analyzed statistically, even though it may be only to calculate the mean and standard deviation. Look at the numbers to check if they are reasonable. When there is an exceptionally good result, either one has made a discovery or, unfortunately, a mistake.

Tensile Strength

With metals and plastics, tensile strength is normally measured with a dog bone sample, where the thin part of the bone is the gage length. All of the material in the gage length is in the maximum stress zone making dog bones the best way to measure tensile strength. There is a catch when dealing with ceramics. Being brittle and having a high modulus of elasticity, they are very difficult to align and obtain a straight pull. Fancy air bearings and special alignment fixtures have been used, but are not common. Two methods exist for measuring tensile strength that do not use dog bones, however. They are the brittle ring test and the hydrostatic test.

Brittle Ring Test [6]

In this test, a disc or ring is placed on the press platen and broken in compression as seen in Figure 11.69.

This test has a couple of nice features. The sample is easy to prepare, and the test setup is very simple, requiring only compression platens. In the reference, results from the brittle ring test are compared with those from both tensile and MOR tests. Comparison is fairly good, but which set of data best reflects the closest fit to the actual tensile strength? Like most tests, when they are used to compare materials of similar type, they are valid within the constraints of the test set.

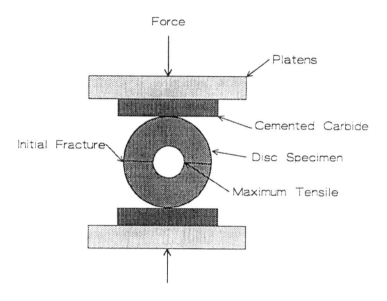

Figure 11.69: Brittle Ring Test, Sketch. Can be instrumented with strain gages for calibration.

Hydrostatic

In this test, the sample is a tube that is stressed by forcing a liquid (water) into the tube under pressure until it breaks. Figure 11.70 is a sketch of the experimental setup.

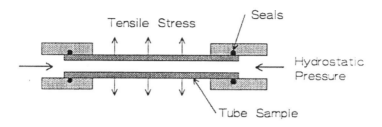

Figure 11.70: Hydrostatic Tensile Test. Yields true tensile strength measurements.

This is an unambiguous tensile test using a sample shape that is sometimes easy to make. Of course, it is only valid for impervious samples or samples with an impervious elastomer liner. Tubes are made by slip casting, injection molding, isopressing, or by extrusion.

Compressive Strength

Compressive strength is measured on cylindrical (round or square) samples with square and parallel ends. Flat and parallel platens are used to compress the sample. Many ceramics are so strong that they will damage steel platens in the test. To avoid this, place the sample on cemented carbide blocks. Perfect alignment is difficult to achieve, so it helps to use the alignment fixture shown in Figure 6.16. A small piece of paper on both ends also helps to smooth out minute irregularities, as the paper compresses during the test and acts as a cushion. There is a lot of elastic energy stored in strong samples and they explode causing a safety hazard for the operator and others in the room. Shield the sample to prevent this hazard.

Modulus of Elasticity

The problem in measuring the modulus of elasticity (MOE) is accurately measuring strain, which can be very small. Extensometers can be used, but must be very sensitive. Low MOE ceramics are less demanding.

Direct Strain Measurements

One type of setup is to contact the bottom of a MOR test bar with a probe that measures linear displacement. Commercial equipment is available and is shown in Figure 11.71.

Figure 11.71: Strain Measurement, Extensometer. Measures strain when attached to the specimen. (Courtesy of MTS)

The four-point-bend fixture is at the top of the support column. There is a central rod touching the bottom of the test bar and transmitting the deflection to the extensometer at the lower right. This method is a direct measure of deflection.

Sonic Measurements

There are two types of sonic measurements: sonic velocity and resonate frequency.

Sonic Velocity. By placing a transducer on one side of the sample and the receiver on the other, the sonic velocity can be measured. Some instruments (V meter, James) have only one probe and use the reflected sonic wave from the other side for the measurement. In order to obtain a true MOE, the velocity of the shear wave must also be determined; this is harder to do. Sonic velocity by itself is a useful and easy measurement. Coupled with a measure of the amplitude attenuation, one obtains a fairly good idea about the nature of the ceramic body. For example, sonic velocity has sufficient sensitivity to discern the pressing direction in a part, as the velocity is higher in the pressing direction than along the sides. Attenuation is an empirical method to estimate how well a body is bonded. This is easily accomplished by measuring the reduction of wave amplitude across the sample. Techniques of this sort are useful quality control tools. They can even pick up interior cracks and laminations, as the reflected wave has traveled only half of the distance that it would have if the piece were whole.

Resonate Frequency. This technique is based on the resonate frequency or ring of a sample as it is tapped while suspended on end supports. A sonic probe is placed in contact with the side of the sample as it is tapped. The MOE measurement appears in a display. Shear modulus can also be measured by relocating the probe, but it is a little more difficult to get a reading. An instrument is shown in Figure 11.72.

Figure 11.72: Resonate Frequency Measurement Apparatus for MOE measurements. Easy to use. (Courtesy of Grindosonic/Lemmings)

The device is commercially available and inexpensive. The test is fast and easy and can be made on any cylindrical sample of sufficient (3-4") length. Bar or cylindrical rod samples are usually tested.

Hardness

Hardness measurements are made by pressing a precision diamond point onto the polished surface of the sample with a certain load, and then measuring the size of the indentation. A special instrument is used for this purpose. Before purchasing it, ensure that the instrument is also capable of measuring fracture toughness. Fracture toughness requires a higher load than micro-hardness.

The load for hardness measurements is between 100-200 grams, as it is not advisable to induce fractures around the indent. In comparison, for fracture toughness, the indent must be fractured and loads of a few kilograms are used.

There are two types of diamond configurations commonly used for ceramics: Vickers and Knoop. Vickers produces a symmetrical indent of pyramidal shape. The Knoop indenter produces an elongated pyramidal shape. Vickers is more commonly used as it is also the shape for fracture toughness measurements. As previously mentioned, it is much easier to see the ends of the indentations when the sample is coated with a reflective metal film such as gold. Now that the ends are clearly visible, the measurement is longer. This results in a lower hardness calculated for an indent without the reflective film. It is to be consistant, one way or the other.

A sketch of an indent will be shown in the next section along with the measurements for fracture toughness.

Fracture Toughness

There are a variety of configurations used on samples for measuring fracture toughness.[7] These are shown redrawn, in Figure 11.73.

In addition to these five configurations, another also exists which will be described shortly. In Figure 11.73 each of the specimen geometries is as follows:

"A". Double Cantilever Beam. It is required that the crack length be followed, making it difficult for opaque materials.

"B". Single Edged Notched Beam. This is similar to four- point bending. Only a initial crack must be formed and presents complications.

"C". Double Torsion. This sketch is shown from the end view. From the top, the specimen is rectangular where the length is 3-4 times the plate width. The initial crack extends in only a short distance. The test is not too difficult to perform.

"D". Chevron Notch. A notch is sawed in the specimen as seen in the shaded area in the end view. This eliminates the need for pre-cracking, which can be tricky. On the other hand, cutting the chevron notch can be difficult to do correctly.

"E". Indentation strength. The sample is indented with a load high enough to initiate cracks out of the indent corners. Then the specimen is broken in bending.

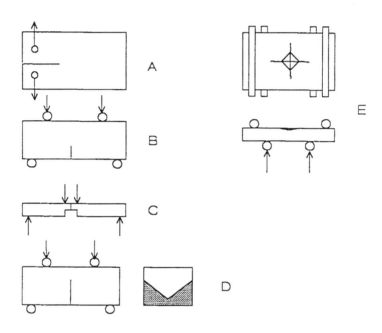

Figure 11.73: Fracture Toughness Specimen Configurations, Sketch. Pick the one that fits your needs.

An additional method that is fairly simple is the "Indentation Crack Length" method where the specimen is indented with a Vickers diamond point at a load high enough to cause fracturing at the corners, as seen in Figure 11.74.

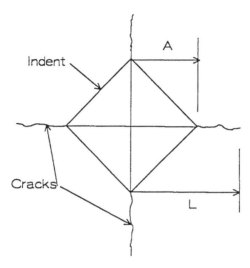

Figure 11.74: Fracture Toughness Test by the Indentation Crack Length Method. Arrows point to the measurements needed.

When working with a fine-grained, dense specimen, the Indentation Crack Length is most commonly used as it is easy to preform the test and make the measurements. However, there are also problems with this method, as shown in Figure 11.75.

The sample has been coated with gold/palladium and is viewed with Nomarsky interference contrast. While the indent and corner cracks are clearly seen, there are also circular cracks around the indent that compromise the test results. In this case, the load should be reduced or another method used to measure fracture toughness.

Figure 11.75: Vickers Indentation with Cracking. Polished section with polarized light and Nomarski interference contrast illumination.Ring cracks compromise the toughness measurement. Scale bar 200 μm.

A simple equation is used to calculate toughness (K_{IC}).

$$K_{IC} = x(E/H)^{1/2} (P/c^{3/2})\qquad\qquad (11.9)$$

Where:

E = Young`s modulus, often taken from the literature

x = a constant

H = the hardness

c = the average crack length

P = the indentation load

Indentation methods are not suitable for coarse or porous materials as a clean indent is not possible. Such materials require a different technique. These are difficult and as such there is very little data regarding this in the literature. Referring back to the reference, the chevron notch, double cantilever, or single-edged, notched beam are possible methods for making this measurement. Special equipment and know-how is needed to perform this correctly, so it is a good idea to call upon an expert to get started.

Two of these experimental setups will be shown. The first is in Figure 11.76.

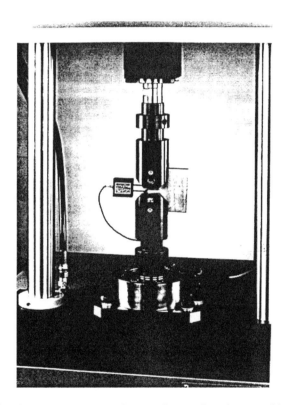

Figure 11.76: Fracture Toughness Setup for the Double Cantilever Configuration. Fixture is commercially available.(Courtesy of Instron)

This is a setup on a test machine for the Double Cantilever Beam. There are two pins that pass through the holes in the specimen; these are used to apply the tensile stress on the pre-crack. The other example is seen in Figure 11.77.

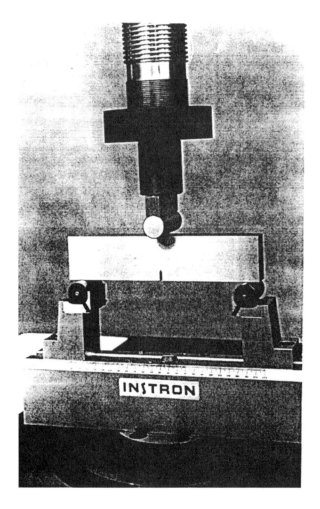

Figure 11.77: Fracture Toughness Setup for the Single-Edged Notched Beam Configuration. (Courtesy of Instron)

The picture is of a Single-Edged, Notched Beam configuration setup for three-point loading. Various-sized beams can be used on the same fixture. Also note that the lower pins are free to roll as the beam increases in length as it bends. Two little springs hold the roller pins in place until the sample is loaded.

Wear

This discussion starts by looking at an example of the complexity of wear.[8] Wear of a high quality TZP sliding on itself in a version of a four ball wear tester is seen in Figure 11.78.

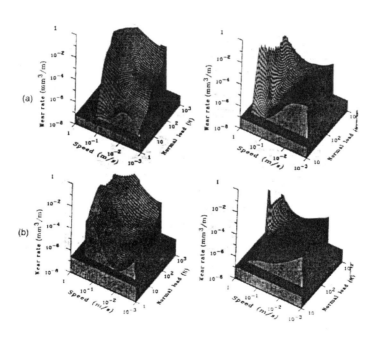

Figure 11.78: Wear Maps, TZP with Different Lubricants. Wear is complex especially when the mass transport mechanism changes with conditions.

There are three independent variables in the test: speed, load, and lubrication conditions. The wear maps are quite different for each set of conditions. The authors rationalized this behavior by calculating a critical velocity with an equation containing material thermal properties and friction, which affect thermal shock in the ceramic. Above that velocity, the surface breaks down and spalls. Wear is then by fracture and displacement. Below that transition, wear behavior is not so clear-cut.

Other examples of wear surfaces on ceramics were seen in the previous section on microscopy and illustrate that, as conditions change, the wear mechanism can change. There is the additional complication of the consequences of introducing other materials. Then, chemical reactions can influence wear rates as well as normal load and velocity.

One helpful way to envision wear, as previously mentioned, is to consider the following mass transport mechanisms: plastic deformation, diffusion, viscous flow, and fracture with kinetic displacement. Any, or all, of these may be involved in the wear process at one point or another. Microscopic examination and surface analysis provide clues to the mass transport mechanism that is active. A brief discussion on the four mass transport mechanisms (wear) appears below.

Plastic Deformation

Surfaces on wear couples are subjected to very severe conditions of pressure, shear, and temperature. Either the materials themselves or reaction products can be transported by plastic deformation along the sliding direction. This proves tricky, as the rate-controlling step can be the chemical reaction step, deluding the investigator into thinking that diffusion is the wear mechanism itself. To take an example, alumina sliding on steel results in a chemical reaction forming a spinel (Al_2O_3-Fe_2O_3). This reaction was first recognized by Coes[9] and later by Brown.[10] While that reaction does displace Al, it does not displace it very far. Something else dominates the wear phenomena. The spinel crystallizes in the cubic system that has sufficient (at least three) slip systems for volumetric plastic deformation. Under the conditions of shear, pressure, and temperature at the interface, the spinel is whisked away as soon as it is formed. Coes found the spinel in the wear detritus, where it is expected. Plastic deformation dominates this wear process. Diffusion (chemical reaction) dominates the wear rate. Thinking of wear in mass transport terms helps to clarify the process.

Diffusion

Compared to the three other wear processes, diffusion is relatively static, with short transport distances. This is not to say that diffusion does not play a significant role in wear processes; it does. A clean example of diffusive wear is corrosion or oxidation. Mass transport from one crystal lattice to another by diffusion is a common phenomenon. One ceramic example is the oxidation of the surface of a $MoSi_2$ heating element. Oxygen diffuses in to form the silica film that protects the element from rapid deterioration. As the film forms, diffusion is slowed by the parabolic rate law where film thickness is inversely proportional to time. Eventually, the film spalls off by stresses caused by differential thermal expansion between the $MoSi_2$ element and the silica film. Mass transport then is by fracture and kinetic displacement.

Viscous Flow

A ceramic example of viscous flow wear is in ferrous contact refractories, such as a teeming zirconia nozzle. It is not the molten steel that wears the nozzle, but the slag inclusions in the steel that wet and adhere to the zirconia surface. At these temperatures, the slag dissolves the zirconia and moves downstream by viscous flow. Solution is an active process that involves diffusion. However, this is not how the predominate mass transport occurs.

Fracture and Kinetic Displacement

While there are many good examples of the fracture and kinetic displacement wear process in ceramics, we will discuss just one example.

A ceramic plant manufacturing $BaO-TiO_2$ for capacitors developed a problem with alumina contamination that was serious as it negatively affects the dielectric properties. Running out of inventory of good product spurred some aggressive action. All of the raw materials were within specification.

Alumina had to be picked up in processing. One of these steps involved milling, using alumina milling media. This was the standard procedure. Small star cracks were observed on the surface of the mill balls. They then discovered that a new rubber lining with lifter bars had been installed in the mill. This caused the balls to be lifted higher than before, cracking them when they crashed down. Minute particles of alumina were fractured and displaced kinetically, contaminating the product.

Wear Tests

A common wear test is the pin on disc, where a rod is pressed against a rotating disc with a known force, at a known sliding speed, and with a lubricant. Figure 11.79 is a sketch of a pin on disc wear apparatus.

Interfacial temperatures can become very high at the points of the aspirates, making the ambient temperature deceptive as a measure of the processes at the interface. There are options on pin on disc test apparatus such as speed, force, temperature, and lubricant. Conditions are selected that most closely match those of the intended service.

A multitude of wear test machines and methods are available, ranging from sandblasting to reciprocating brushes.

While some wear tests are potentially useful for screening materials, there is no substitute for testing in the actual application. This requires a close relationship between the supplier of the wear parts and the potential customer. Liability can be a problem unless the user has a test facility for evaluating parts. There is an understandable reticence in placing experimental parts into commercial installations. This is especially true when failure of the part would result in a disaster, such as spilling a ladle of molten steel on the mill floor.

In one case, a well-conceived composition, carefully crafted, and thoroughly-lab-tested refractory never made it into the application.[11] Later, a poorly conceived, shop-made, and scantily tested refractory made the installation.

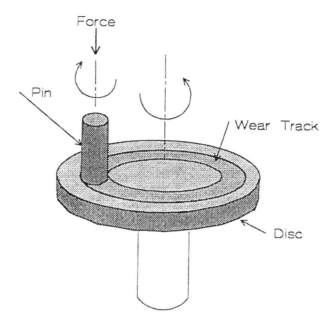

Figure 11.79: Pin and Disc Wear Test Configuration. A common wear test for evaluating materials.

Thermal Shock

Thermal shock is measured by heating the specimens to a temperature and then quenching them in water.[12] Residual strength is measured as a function of the temperature differential between the specimens and the water. The typical response is shown in Figure 11.80.

Residual strength is unchanged up to a point and then suddenly plunges down to a lower value. It then flattens out for an interval and slowly drops off. The parameters of interest are the temperature differentials where the sudden drop occurs and how far the strength drops down. Curve "B" is better in thermal shock than curve "A". Constructing this diagram is a bit labor-intensive. For example, if there are four replicate test bars for each temperature and five temperatures, this results in 20 specimens that must be heated, shocked, and MOR-measured. It should be noted, however, that this procedure is derived from thermal shock theory and is valid. Other methods are empirical, which may be useful when simulating actual conditions in service. There are many of these, including an acetylene torch traverse, cyclic torch tests, placing the specimen over a burner, plunging into a molten metal, and inserting a specimen into a hot kiln, to name a few.

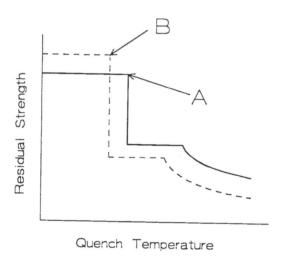

Figure 11.80: Thermal Shock, Strength Response. The two measurements of interest are shown in the figure.

Pycnometer

A pycnometer can be used for density measurements. The sample volume is obtained by the amount of gas displaced in a fixed volume. This, along with the sample weight, is used to measure density.

Porosity Volume and Size Distribution

A porosimeter is used for measuring the pore volume and pore size distribution. Mercury is forced into the pores, with the pressure and displaced-pore volume measurements used to calculate the size distribution. This is a useful technique for looking at the early stages of sintering and porous materials. Since mercury does not wet the ceramic, higher pressures are required to infiltrate the smaller pores.

Mercury has a vapor pressure and is toxic over a period of time. Spills are inevitable and can be difficult to clean. Mercury can also be ingested by contaminating food or tobacco carelessly placed around the porosimeter. A mercury porosimeter should be isolated, with venting and a means to contain spills. Some mercury is left in the pores after the test, creating a sample disposal problem.

Check List, Physical Properties

- MOR
  - Bar size
  - Surface finish
  - Four point bending
  - Statistical analysis
  - Examination of data
- Tensile tests
  - Dog bones are difficult, alignment
  - Brittle ring test
  - Hydrostatic, true values
- Compressive
  - Square and parallel
  - Alignment fixture
  - Paper cushion
  - Safety shield
- MOE
  - Direct strain measurement
  - Machine compliance

Sonic velocity
    Easy
    Useful
    Amplitude attenuation
Resonate Frequency
    Easy
- Hardness
    Polished surfaces
    Vicker's
    Measure diagonals
- Fracture toughness
    Fine-grained materials
        Methods
        Vicker's, high load
        Measure crack length and indent diagonal
        Calculations
    Coarse-grained materials
        Other methods
        Consult an expert
- Wear resistance
    Wear maps
    Mass transport mechanisms
    Variable conditions
    Interactions between variables
    Pin on disc
    Use actual application
- Thermal shock
    Quenching
    MOR measurements
    Curve critical parameters

REFERENCES

1. Engineering Property Data on Selected Ceramics. Vol. III, Single Oxides. Battelle, July 1981.
2. Robert Samuel and Srinivasan Chandrasekar, "Effect of Residual Stresses on the Fracture of Ground Ceramics," *J. Am. Ceram. Soc.* **72**[10] 1960-66 (1989).
3. A. G. King, "Tribology of Abrasive Machining," *Thin Solid Films* 108, 127-34 (1983).
4. Alan G. King and W. M. Wheildon, Ceramics in Machining Processes: New York: Academic Press, 1966.
5. A. H. Heuer, "Alloy Design in Partially Stabilized Zirconia," Science and Technology of Zirconia, Edited by A. H. Heuer and L. W. Hobbs. Westerville, Ohio: The American Ceramic Society, 1981, pp. 98-115.
6. S. A. Bortz and H. H. Lund,. "The Brittle Ring Test," Mechanical Properties of Engineering Ceramics, Edited by W. Wurth Kriegel and Hayne Palmour III. New York: Interscience Publishers, 1961, pp. 383-406.
7. Richard M. Anderson,. "Testing Advanced Ceramics," *Advanced Materials and Processes* (3/89).
8. Soo E. Lee, Stephen M. Hsu, and Ming C. Shen, "Ceramic Wear Maps: Zirconia," *J. Am. Ceram. Soc.* **76**[8] 1937-47 (1993).
9. L. Coes Jr., "Chemistry of Abrasive Action," *Ind. Eng. Chem.* 47, 2493-4 (1955).
10. W. R. Brown, N. S. Eiss, H. T. McAdams, "Chemical Mechanisms Contributing to Wear of Single Crystal Sapphire on Steel," *J. Am. Ceram. Soc.* **47** 157-62 (1964).
11. D. F. Beal, Personal Communication.
12. D. P. H. Hasselman, "Unified Theory of Thermal Shock Fracture Initiation and Crack Propagation in Brittle Ceramics," *J. Am. Ceram. Soc.* **52**[11], 600-04 (1969).

12

Tools

1.0 INTRODUCTION

Careful statistical analysis of an exhaustive study revealed that one spends more time looking for tools than using them. Since there is a tendency to misplace tools, a laboratory may lack tools rather than have an abundance of them. Periodic search-and-seize missions throughout the lab are as gratifying as they are deeply resented. In the unlikely instance that one finds the tools, they are usually coated with dried slip, rusted, broken, worn, or dull.

2.0 TOOL SECURITY

It is axiomatic that when one reaches for a tool, it is not available. While color-coding with distinguishing paints helps maintain identifiable tools, it is usually not enough. Since tool security is often an issue, one might find the following suggestions helpful.
- Lock the tools and carry the key in your pocket.
- Have a duplicate set for personal use.
- Hide them where it is unlikely that they will be found.

3.0 CRAFTSMAN HAND TOOLS

One commonly finds the following hand tools in a tool box: screw drivers, wrenches, hammers, pliers, cutters, and others. One must acquire a variety of tools and in different sizes. Additionally, one should check the quality of the manufacture, especially the quality of the steel. Cheap tools are made from soft steel. These are cheaper to make but more costly to use as they wear out quickly and do not work well. The round-ended screw driver is a common apparition.

Locked tool cabinets with the identity and location of each tool is a good way to store hand tools.

4.0 MEASURING HAND TOOLS

An initial question centers on the measuring system. Some people believe that the SI system is universal; however, there is a problem. Most laboratories have a collection of old and new equipment, with the old calibrated in English units and the new calibrated in either English or metric. SI units are not likely except where they coincide with metric. The bolts, screws, gears, slots, notches and shims are all in English units. The author has no problem with the SI or the metric systems as inherently more reasonable, based on decades of ten and standardization of fundamental units. However, at issue here is the everyday function of the laboratory that is predominantly English.

Geometrical

There are sophisticated measurement systems for almost everything. Of concern here are those hand-measuring tools kept in the tool cabinet or drawer.

Linear

Micrometers often have carbide faces that grab and give false readings. Carbide faces wear at a slower rate than steel. Steel faces have a better feel and do not wear fast. Avoid vernier calipers that have tips ground down to a knife edge; these wear very quickly, especially with abrasive ceramics.

Angular

Angle measuring spirit levels should be rugged to sustain lab use. Accurate angular measurements are made in the shop with dial gages, sine bar, and gage blocks. Go to the shop for these measurements.

Mass

Digital scales have nearly displaced analog types, because they are so easy to use. The tare button is one of civilization's greatest inventions. Keep scales in calibration.

Temperature

When the indicator fluid separates, coalesce the fluid by chilling the thermometer to contain all the fluid in the bulb and then warm it again. Thermocouples or optical pyrometers are common for temperature measurements; these should be often calibrated against a traceable standard. As a backup, pyrometric cones are still useful as they also give a qualitative estimate of the total heat treatment.

Pressure/Vacuum

Bourdon-type gages are often used to measure pressure and vacuum. These gages can become plugged with slip or other material and, when overloaded, give false readings. It is wise to often calibrate them against a standard gage. Glycerin-filled gages are very useful for vibrating apparatus, like a hydraulic press. Vacuum gages are sometimes calibrated in "inches of mercury" which is a holdover from the beginning of the technology. The early instruments were of the barometer type using a mercury column and a scale.

Electrical

In the laboratory, a multimeter receives much use. Hooking them up incorrectly results in their failure and need to be replaced.

Dedicated Tools

When a particular piece of equipment requires a special tool for use, one should fasten the tool to the equipment with a chain. Such an example is with the pressure infiltration apparatus discussed earlier. The snap ring had to be removed after each run to clean the valve. This requires special pliers fastened to the apparatus with a chain.

5.0 POWER TOOLS

In the laboratory, one only needs a few power tools. A power drill and a small, high-speed grinder are useful. Off-site locations sometimes require battery-powered tools. Remember to keep them charged.

6.0 MACHINE TOOLS

A laboratory working with the development of structural ceramics may have to get a surface grinder for making test bars. Problematic is the lead time from doing the experiment and obtaining the results. The difference can range from two days to four weeks. If bureaucratically possible, one can attain efficiency by eliminating the chance of a bottleneck.

Check List, Tools

- Security
     Accept failure
- Hand Tools
     Good steel
- Measuring Tools
     Units of measurements
     Geometrical
     Micrometers/calipers
- Electronic Balances
- Temperature
     Calibrate
- Pressure/Vacuum
     Calibrate
     Glycerin
- Dedicated Tools
     Chain
- Machine Tools
     Break any bottleneck.

Index